W9-BYK-208

Getting Started with Processing

SECOND EDITION

Casey Reas and Ben Fry

MAKER MEDIA™
SAN FRANCISCO, CA

Advance Praise for *Getting Started with Processing*

"Making a computer program used to be as easy as turning it on and typing one or two lines of code to get it to say, *Hello.* Now it takes a 500+-page manual and an entire village. Not anymore. This little book by Ben and Casey gets you computationally drawing lines, triangles, and circles within minutes of clicking the *download* button. They've made making computer programs humanly and humanely possible again—and that's no small feat."

— *John Maeda, President of Rhode Island School of Design*

"*Getting Started with Processing* is not only a straightforward introduction to basic programming—it's fun! It almost feels like an activity workbook for grownups. You may want to buy it even if you never thought you were interested in programming, because you will be."

— *Mark Allen, Founder and Director, Machine Project*

"This is an excellent primer for those wanting to dip their feet into programming graphics. Its learning by doing approach makes it particularly appropriate for artists and designers who are often put off by more traditional theory-first approaches. The price of the book and the fact that the Processing environment is open source makes this an excellent choice for students."

— *Gillian Crampton Smith, Fondazione Venezia Professor of Design, IUAV University of Venice*

Make: Getting Started with Processing

by Casey Reas and Ben Fry

Published by Maker Media, Inc., 1160 Battery Street East, Suite 125, San Francisco, CA 94111.

Maker Media books may be purchased for educational, business, or sales promotional use. Online editions are also available for most titles (http://safaribooksonline.com). For more information, contact our corporate/institutional sales department: 800-998-9938 or corporate@oreilly.com.

Editor: Anna Kaziunas France	**Indexer:** Wendy Catalano
Production Editor: Nicole Shelby	**Interior Designer:** David Futato
Copyeditor: Jasmine Kwityn	**Cover Designer:** Casey Reas
Proofreader: Kim Cofer	**Illustrator:** Rebecca Demarest

September 2015:   Second Edition

Revision History for the Second Edition

2015-09-03:   First Release

See http://oreilly.com/catalog/errata.csp?isbn=9781457187087 for release details.

978-1-457-18708-7

[LSI]

Contents

Preface

We created Processing to make programming interactive graphics easier. We were frustrated with how difficult it was to write this type of software with the programming languages we usually used (Java and C++), and were inspired by how simple it was to write interesting programs with the languages of our childhood (Logo and BASIC). We were most influenced by Design By Numbers (DBN), a language we were maintaining and teaching at the time (and which was created by our research advisor, John Maeda).

Processing was born in spring 2001 as a brainstorming session on a sheet of paper. Our goal was to make a way to sketch (prototype) the type of software we were working on, which was almost always full-screen and interactive. We were searching for a better way to test our ideas easily in code, rather than just talking about them or spending too much time programming them in C++. Our other goal was to make a language for teaching design and art students how to program and to give more technical students an easier way to work with graphics. The combination is a positive departure from the way programming is usually taught. We begin by focusing on graphics and interaction rather than on data structures and text console output.

Processing experienced a long childhood; it was alpha software from August 2002 to April 2005 and then public beta software until November 2008. During this time, it was used continuously in classrooms and by thousands of people around the world. The language, software environment, and curricula around the project were revised continuously during this time. Many of our original decisions about the language were reinforced and many were changed. We developed a system of software extensions, called *libraries*, that have allowed people to expand Processing into many unforeseen and amazing directions. (There are now over 100 libraries.)

In fall 2008, we launched the 1.0 version of the software. After seven years of work, the 1.0 launch signified stability for the language. We launched the 2.0 release in spring 2013 to make the software faster. The 2.0 releases introduced better OpenGL integration, GLSL shaders, and faster video playback with GStreamer. The 3.0 releases in 2015 make programming in Processing easier with a new interface and error checking while programming.

Now, fourteen years after its origin, Processing has grown beyond its original goals, and we've learned how it can be useful in other contexts. Accordingly, this book is written for a new audience—casual programmers, hobbyists, and anyone who wants to explore what Processing can do without getting lost in the details of a huge textbook. We hope you'll have fun and be inspired to continue programming. This book is just the start.

While we (Casey and Ben) have been guiding the Processing ship through the waters for the last twelve years, we can't overstate that Processing is a community effort. From writing libraries that extend the software to posting code online and helping others learn, the community of people who use Processing has pushed it far beyond its initial conception. Without this group effort, Processing would not be what it is today.

How This Book Is Organized

The chapters in this book are organized as follows:

- Chapter 1: Learn about Processing.
- Chapter 2: Create your first Processing program.
- Chapter 3: Define and draw simple shapes.
- Chapter 4: Store, modify, and reuse data.
- Chapter 5: Control and influence programs with the mouse and the keyboard.
- Chapter 6: Transform the coordinates.
- Chapter 7: Load and display media including images, fonts, and vector files.
- Chapter 8: Move and choreograph shapes.
- Chapter 9: Build new code modules.
- Chapter 10: Create code modules that combine variables and functions.
- Chapter 11: Simplify working with lists of variables.
- Chapter 12: Load and visualize data.
- Chapter 13: Learn about 3D, PDF export, computer vision, and reading data from an Arduino board.

Who This Book Is For

This book is written for people who want a casual and concise introduction to computer programming, who want to create images and simple interactive programs. It's for people who want a jump-start on understanding the thousands of free Processing code examples and reference materials available online. *Getting Started with Processing* is not a programming textbook; as the title suggests, it will get you started. It's for teenagers, hobbyists, grandparents, and everyone in between.

This book is also appropriate for people with programming experience who want to learn the basics of interactive computer graphics. *Getting Started with Processing* contains techniques that can be applied to creating games, animation, and interfaces.

Conventions Used in This Book

The following typographical conventions are used in this book:

Italic
: Indicates new terms, URLs, email addresses, filenames, and file extensions.

`Constant width`
: Used for program listings, as well as within paragraphs to refer to program elements such as variable or function names, databases, data types, environment variables, statements, and keywords.

`Constant width italic`
: Shows text that should be replaced with user-supplied values or by values determined by context.

 This element signifies a tip, suggestion, or general note.

 This element indicates a warning or caution.

Using Code Examples

This book is here to help you get your job done. In general, you may use the code in this book in your programs and documentation. You do not need to contact us for permission unless you're reproducing a significant portion of the code. For example, writing a program that uses several chunks of code from this book does not require permission. Selling or distributing a CD-ROM of examples from Make: books does require permission. Answering a question by citing this book and quoting example code does not require permission. Incorporating a significant amount of example code from this book into your product's documentation does require permission.

We appreciate, but do not require, attribution. An attribution usually includes the title, author, publisher, and ISBN. For example: *"Getting Started with Processing* by Casey Reas and Ben Fry. Copyright 2015 Casey Reas and Ben Fry, 978-1-457-18708-7."

If you feel your use of code examples falls outside fair use or the permission given here, feel free to contact us at *bookpermissions@makermedia.com*.

Safari® Books Online

Safari Books Online is an on-demand digital library that delivers expert content in both book and video form from the world's leading authors in technology and business.

Technology professionals, software developers, web designers, and business and creative professionals use Safari Books Online as their primary resource for research, problem solving, learning, and certification training.

Safari Books Online offers a range of plans and pricing for enterprise, government, education, and individuals.

Members have access to thousands of books, training videos, and prepublication manuscripts in one fully searchable database from publishers like Maker Media, O'Reilly Media, Prentice Hall Professional, Addison-Wesley Professional, Microsoft Press, Sams, Que, Peachpit Press, Focal Press, Cisco Press, John Wiley & Sons, Syngress, Morgan Kaufmann, IBM Redbooks, Packt, Adobe Press, FT Press, Apress, Manning, New Riders, McGraw-Hill, Jones & Bartlett, Course Technology, and hundreds more. For more information about Safari Books Online, please visit us online.

How to Contact Us

Please address comments and questions concerning this book to the publisher:

Maker Media, Inc.
1160 Battery Street East, Suite 125
San Francisco, California 94111
800-998-9938 (in the United States or Canada)
http://makermedia.com/contact-us/

Make: unites, inspires, informs, and entertains a growing community of resourceful people who undertake amazing projects in their backyards, basements, and garages. Make: celebrates your right to tweak, hack, and bend any technology to your will. The Make: audience continues to be a growing culture and community that believes in bettering ourselves, our environment, our educational system—our entire world. This is much more than an audience, it's a worldwide movement that Make: is leading—we call it the Maker Movement.

For more information about Make:, visit us online:

Make: magazine: *http://makezine.com/magazine/*
Maker Faire: *http://makerfaire.com*
Makezine.com: *http://makezine.com*
Maker Shed: *http://makershed.com/*

We have a web page for this book, where we list errata, examples, and any additional information. You can access this page at: *http://shop.oreilly.com/product/0636920031406.do*

To comment or ask technical questions about this book, send email to *bookquestions@oreilly.com*.

Acknowledgments

For the first and second editions of this book, we thank Brian Jepson for his great energy, support, and insight. For the first edition, Nancy Kotary, Rachel Monaghan, and Sumita Mukherji gracefully carried the book to the finish line. Tom Sgouros made a thorough edit of the book and David Humphrey provided an insightful technical review.

We can't imagine this book without Massimo Banzi's *Getting Started with Arduino* (Maker Media). Massimo's excellent book is the prototype.

A small group of individuals has, for years, contributed essential time and energy to Processing. Dan Shiffman is our partner in the Processing Foundation, the 501(c)(3) organization that supports the Processing software. Much of the core code for Processing 2.0 and 3.0 has come from the sharp minds of Andres Colubri and Manindra Moharana. Scott Murray, Jamie Kosoy, and Jon Gacnik have built a wonderful web infrastructure for the project. James Grady is rocking the 3.0 user interface. We thank Florian Jenett for his years of diverse work on the project including the forums, website, and design. Elie Zananiri and Andreas Schlegel have created the infrastructure for building and documenting contributed libraries and have spent countless hours curating the lists. Many others have contributed significantly to the project; the precise data is available at *https://github.com/processing*.

The Processing 1.0 release was supported by Miami University and Oblong Industries. The Armstrong Institute for Interactive Media Studies at Miami University funded the Oxford Project, a series of Processing development workshops. These workshops were made possible through the hard work of Ira Greenberg. These four-day meetings in Oxford, Ohio, and Pittsburgh, Pennsylvania, enabled the November 2008 launch of Processing 1.0. Oblong Industries funded Ben Fry to develop Processing during summer 2008; this was essential to the release.

The Processing 2.0 release was facilitated by a development workshop sponsored by New York University's Interactive Telecommunication Program. The work on Processing 3.0 was generously sponsored by the Emergent Digital Practices program at the University of Denver. We thank Christopher Colemen and Laleh Mehran for the essential support.

This book grew out of teaching with Processing at UCLA. Chandler McWilliams has been instrumental in defining these classes. Casey thanks the undergraduate students in the Department of Design Media Arts at UCLA for their energy and enthusiasm. His teaching assistants have been great collaborators in defining how Processing is taught. Hats off to Tatsuya Saito, John Houck, Tyler Adams, Aaron Siegel, Casey Alt, Andres Colubri, Michael Kontopoulos, David Elliot, Christo Allegra, Pete Hawkes, and Lauren McCarthy.

Through founding the Aesthetics and Computation Group (1996–2002) at the MIT Media Lab, John Maeda made all of this possible.

1/Hello

Processing is for writing software to make images, animations, and interactions. The idea is to write a single line of code, and have a circle show up on the screen. Add a few more lines of code, and the circle follows the mouse. Another line of code, and the circle changes color when the mouse is pressed. We call this *sketching* with code. You write one line, then add another, then another, and so on. The result is a program created one piece at a time.

Programming courses typically focus on structure and theory first. Anything visual—an interface, an animation—is considered a dessert to be enjoyed only after finishing your vegetables, usually several weeks of studying algorithms and methods. Over the years, we've watched many friends try to take such courses and drop out after the first lecture or after a long, frustrating night before the first assignment deadline. What initial curiosity they had about making the computer work for them was lost because they couldn't see a path from what they had to learn first to what they wanted to create.

Processing offers a way to learn programming through creating interactive graphics. There are many possible ways to teach coding, but students often find encouragement and motivation in immediate visual feedback. Processing's capacity for providing that feedback has made it a popular way to approach programming, and its emphasis on images, sketching, and community is discussed in the next few pages.

Sketching and Prototyping

Sketching is a way of thinking; it's playful and quick. The basic goal is to explore many ideas in a short amount of time. In our own work, we usually start by sketching on paper and then moving the results into code. Ideas for animation and interactions are usually sketched as storyboards with notations. After making some software sketches, the best ideas are selected and combined into prototypes (Figure 1-1). It's a cyclical process of making, testing, and improving that moves back and forth between paper and screen.

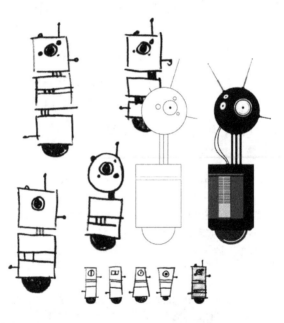

Figure 1-1. *As drawings move from sketchbook to screen, new possibilities emerge*

Flexibility

Like a software utility belt, Processing consists of many tools that work together in different combinations. As a result, it can be used for quick hacks or for in-depth research. Because a Processing program can be as short as one line or as long as thousands, there's room for growth and variation. More than 100 libraries extend Processing even further into domains including sound, computer vision, and digital fabrication (Figure 1-2).

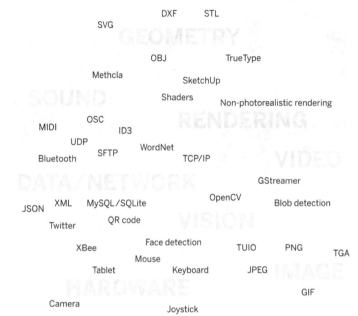

Figure 1-2. *Many types of information can flow in and out of Processing*

Giants

People have been making pictures with computers since the 1960s, and there's much to be learned from this history. For example, before computers could display to CRT or LCD screens, huge plotter machines (Figure 1-3) were used to draw images. In life, we all stand on the shoulders of giants, and the titans for Processing include thinkers from design, computer graphics, art, architecture, statistics, and the spaces between. Have a look at Ivan Sutherland's *Sketchpad* (1963), Alan Kay's *Dynabook* (1968), and the many artists featured in Ruth Leavitt's *Artist and Computer* (Harmony Books, 1976) (*http://www.atariarchives.org/artist/*). The ACM SIGGRAPH and Ars

Electronica archives provide fascinating glimpses into the history of graphics and software.

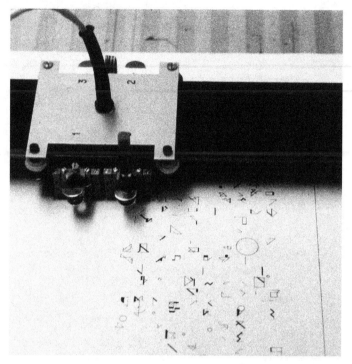

Figure 1-3. *Drawing demonstration by Manfred Mohr at Musée d'Art Moderne de la Ville de Paris using the Benson plotter and a digital computer on May 11, 1971. (photo by Rainer Mürle, courtesy bitforms gallery, New York)*

Family Tree

Like human languages, programming languages belong to families of related languages. Processing is a dialect of a programming language called Java; the language syntax is almost identical, but Processing adds custom features related to graphics and interaction (Figure 1-4). The graphic elements of Processing are related to PostScript (a foundation of PDF) and OpenGL (a 3D graphics specification). Because of these shared features, learning Processing is an entry-level step to programming in other languages and using different software tools.

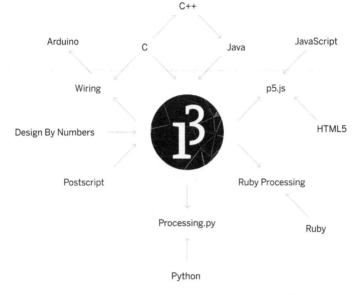

Figure 1-4. *Processing has a large family of related languages and programming environments*

Join In

Thousands of people use Processing every day. Like them, you can download Processing without cost. You even have the option to modify the Processing code to suit your needs. Processing is a *FLOSS* project (that is, *free/libre/open source software*), and in the spirit of community, we encourage you to participate by sharing your projects and knowledge online at Processing.org and at the many social networking sites that host Processing content. These sites are linked from the Processing.org website.

2/Starting to Code

To get the most out of this book, you need to do more than just read the words. You need to experiment and practice. You can't learn to code just by reading about it—you need to do it. To get started, download Processing and make your first sketch.

Start by visiting *http://processing.org/download* and selecting the Mac, Windows, or Linux version, depending on what machine you have. Installation on each machine is straightforward:

- On Windows, you'll have a *.zip* file. Double-click it, and drag the folder inside to a location on your hard disk. It could be *Program Files* or simply the desktop, but the important thing is for the *processing* folder to be pulled out of that *.zip* file. Then double-click *processing.exe* to start.

- The Mac OS X version is a *.zip* file. Double-click it, and drag the Processing icon to the *Applications* folder. If you're using someone else's machine and can't modify the *Applications* folder, just drag the application to the desktop. Then double-click the Processing icon to start.

- The Linux version is a *.tar.gz* file, which should be familiar to most Linux users. Download the file to your home directory, then open a terminal window, and type:

  ```
  tar xvfz processing-xxxx.tgz
  ```

(Replace *xxxx* with the rest of the file's name, which is the version number.) This will create a folder named *processing-3.0* or something similar. Then change to that directory:

```
cd processing-xxxx
```

and run it:

```
./processing
```

With any luck, the main Processing window will now be visible (Figure 2-1). Everyone's setup is different, so if the program didn't start, or you're otherwise stuck, visit the troubleshooting page for possible solutions (*http://bit.ly/process-wiki*).

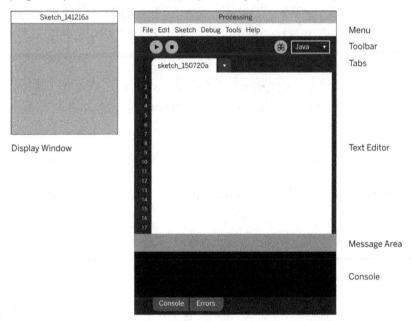

Figure 2-1. *The Processing Development Environment*

Your First Program

You're now running the Processing Development Environment (or PDE). There's not much to it; the large area is the Text Editor, and there's two buttons across the top; this is the Toolbar. Below the editor is the Message Area, and below that is the Console. The Message Area is used for one-line messages, and the Console is used for more technical details.

Example 2-1: Draw an Ellipse

In the editor, type the following:

```
ellipse(50, 50, 80, 80);
```

This line of code means "draw an ellipse, with the center 50 pixels over from the left and 50 pixels down from the top, with a width and height of 80 pixels." Click the Run button the (triangle button in the Toolbar).

If you've typed everything correctly, you'll see a circle on your screen. If you didn't type it correctly, the Message Area will turn red and complain about an error. If this happens, make sure that you've copied the example code exactly: the numbers should be contained within parentheses and have commas between each of them, and the line should end with a semicolon.

One of the most difficult things about getting started with programming is that you have to be very specific about the syntax. The Processing software isn't always smart enough to know what you mean, and can be quite fussy about the placement of punctuation. You'll get used to it with a little practice.

Next, we'll skip ahead to a sketch that's a little more exciting.

Example 2-2: Make Circles

Delete the text from the last example, and try this one:

```
void setup() {
  size(480, 120);
}

void draw() {
  if (mousePressed) {
    fill(0);
  } else {
```

```
    fill(255);
  }
  ellipse(mouseX, mouseY, 80, 80);
}
```

This program creates a window that is 480 pixels wide and 120 pixels high, and then starts drawing white circles at the position of the mouse. When a mouse button is pressed, the circle color changes to black. We'll explain more about this program later. For now, run the code, move the mouse, and click to see what it does. While the sketch is running, the Run button will change to a square "stop" icon, which you can click to halt the sketch.

Show

If you don't want to use the buttons, you can always use the Sketch menu, which reveals the shortcut Ctrl-R (or Cmd-R on the Mac) for Run. The Present option clears the rest of the screen when the program is run to present the sketch all by itself. You can also use Present from the Toolbar by holding down the Shift key as you click the Run button. See Figure 2-2.

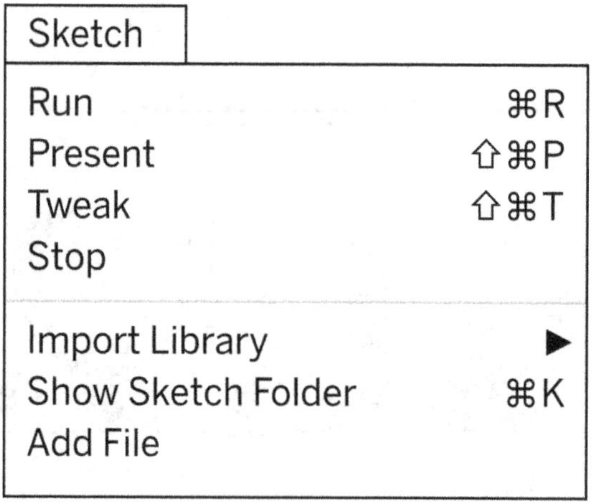

Figure 2-2. *A Processing sketch is displayed on screen with Run and Present. The Present option clears the entire screen before running the code for a cleaner presentation.*

Save and New

The next command that's important is Save. You can find it under the File menu. By default, your programs are saved to the "sketchbook," which is a folder that collects your programs for easy access. Select the Sketchbook option in the File menu to bring up a list of all the sketches in your sketchbook.

It's always a good idea to save your sketches often. As you try different things, keep saving with different names, so that you can always go back to an earlier version. This is especially helpful if—no, *when*—something breaks. You can also see where the sketch is located on your computer with the Show Sketch Folder command under the Sketch menu.

You can create a new sketch by selecting the New option from the File menu. This will create a new sketch in its own window.

Share

Processing sketches are made to be shared. The Export Application option in the File menu will bundle your code into a single folder. Export Application creates an application for your choice of Mac, Windows, and/or Linux. This is an easy way to make self-contained, double-clickable versions of your projects that can run full screen or in a window.

The application folders are erased and re-created each time you use the Export Application command, so be sure to move the folder elsewhere if you do not want it to be erased with the next export.

Examples and Reference

Learning how to program involves exploring lots of code: running, altering, breaking, and enhancing it until you have reshaped it into something new. With this in mind, the Processing software download includes dozens of examples that demonstrate different features of the software.

To open an *example*, select Examples from the File menu and double-click an example's name to open it. The examples are grouped into categories based on their function, such as Form, Motion, and Image. Find an interesting topic in the list and try an example.

All of the examples in this book can be downloaded and run from the Processing Development Environment. Open the examples through the File menu, then click Add Examples to open the list of example packages available to download. Select the *Getting Started with Processing* package and click Install to download.

When looking at code in the editor, you'll see that functions like `ellipse()` and `fill()` have a different color from the rest of the text. If you see a function that you're unfamiliar with, select the text, and then click "Find in Reference" from the Help menu. You can also right-click the text (or Ctrl-click on a Mac) and choose "Find in Reference" from the menu that appears. This will open a web browser and show the reference for that function. In addition, you can view the full documentation for the software by selecting Reference from the Help menu.

The *Processing Reference* explains every code element with a description and examples. The *Reference* programs are much shorter (usually four or five lines) and easier to follow than the longer code found in the *Examples* folder. We recommend keeping the *Reference* open while you're reading this book and while you're programming. It can be navigated by topic or alphabetically; sometimes it's fastest to do a text search within your browser window.

The *Reference* was written with the beginner in mind; we hope that we've made it clear and understandable. We're grateful to the many people who've spotted errors over the years and reported them. If you think you can improve a reference entry or you find a mistake, please let us know by clicking the link at the top of each reference page.

3/Draw

At first, drawing on a computer screen is like working on graph paper. It starts as a careful technical procedure, but as new concepts are introduced, drawing simple shapes with software expands into animation and interaction. Before we make this jump, we need to start at the beginning.

A computer screen is a grid of light elements called *pixels*. Each pixel has a position within the grid defined by coordinates. In Processing, the *x* coordinate is the distance from the left edge of the Display Window and the *y* coordinate is the distance from the top edge. We write coordinates of a pixel like this: (x, y). So, if the screen is 200×200 pixels, the upper-left is (0, 0), the center is at (100, 100), and the lower-right is (199, 199). These numbers may seem confusing; why do we go from 0 to 199 instead of 1 to 200? The answer is that in code, we usually count from 0 because it's easier for calculations that we'll get into later.

The Display Window

The Display Window is created and images are drawn inside through code elements called *functions*. Functions are the basic building blocks of a Processing program. The behavior of a function is defined by its *parameters*. For example, almost every Processing program has a `size()` function to set the width and height of the Display Window. (If your program doesn't have a `size()` function, the dimension is set to 100×100 pixels.)

Example 3-1: Draw a Window

The `size()` function has two parameters: the first sets the width of the window and the second sets the height. To draw a window that is 800 pixels wide and 600 high, type:

```
size(800, 600);
```

Run this line of code to see the result. Put in different values to see what's possible. Try very small numbers and numbers larger than your screen.

Example 3-2: Draw a Point

To set the color of a single pixel within the Display Window, we use the `point()` function. It has two parameters that define a position: the *x* coordinate followed by the *y* coordinate. To draw a little window and a point at the center of the screen, coordinate (240, 60), type:

```
size(480, 120);
point(240, 60);
```

Try to write a program that puts a point at each corner of the Display Window and one in the center. Try placing points side by side to make horizontal, vertical, and diagonal lines.

Basic Shapes

Processing includes a group of functions to draw basic shapes (see Figure 3-1). Simple shapes like lines can be combined to create more complex forms like a leaf or a face.

To draw a single line, we need four parameters: two for the starting location and two for the end.

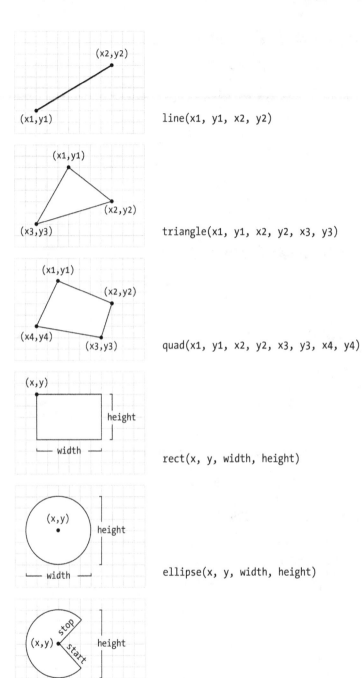

line(x1, y1, x2, y2)

triangle(x1, y1, x2, y2, x3, y3)

quad(x1, y1, x2, y2, x3, y3, x4, y4)

rect(x, y, width, height)

ellipse(x, y, width, height)

arc(x, y, width, height, start, stop)

Figure 3-1. *Shapes and their coordinates*

Example 3-3: Draw a Line

To draw a line between coordinate (20, 50) and (420, 110), try:

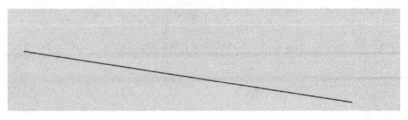

```
size(480, 120);
line(20, 50, 420, 110);
```

Example 3-4: Draw Basic Shapes

Following this pattern, a triangle needs six parameters and a quadrilateral needs eight (one pair for each point):

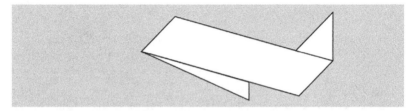

```
size(480, 120);
quad(158, 55, 199, 14, 392, 66, 351, 107);
triangle(347, 54, 392, 9, 392, 66);
triangle(158, 55, 290, 91, 290, 112);
```

Example 3-5: Draw a Rectangle

Rectangles and ellipses are both defined with four parameters: the first and second are for the x and y coordinates of the anchor point, the third for the width, and the fourth for the height. To make a rectangle at coordinate (180, 60) with a width of 220 pixels and height of 40, use the rect() function like this:

```
size(480, 120);
rect(180, 60, 220, 40);
```

Example 3-6: Draw an Ellipse

The x and y coordinates for a rectangle are the upper-left corner, but for an ellipse they are the center of the shape. In this example, notice that the y coordinate for the first ellipse is outside the window. Objects can be drawn partially (or entirely) out of the window without an error:

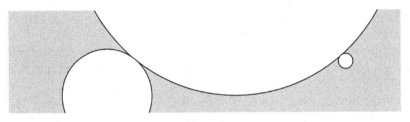

```
size(480, 120);
ellipse(278, -100, 400, 400);
ellipse(120, 100, 110, 110);
ellipse(412, 60, 18, 18);
```

Processing doesn't have separate functions to make squares and circles. To make these shapes, use the same value for the width and the height parameters to ellipse() and rect().

Example 3-7: Draw Part of an Ellipse

The arc() function draws a piece of an ellipse:

```
size(480, 120);
arc(90, 60, 80, 80, 0, HALF_PI);
arc(190, 60, 80, 80, 0, PI+HALF_PI);
arc(290, 60, 80, 80, PI, TWO_PI+HALF_PI);
arc(390, 60, 80, 80, QUARTER_PI, PI+QUARTER_PI);
```

The first and second parameters set the location, the third and fourth set the width and height. The fifth parameter sets the angle to start the arc, and the sixth sets the angle to stop. The angles are set in radians, rather than degrees. Radians are angle measurements based on the value of pi (3.14159). Figure 3-2 shows how the two relate. As featured in this example, four radian values are used so frequently that special names for them were added as a part of Processing. The values PI, QUAR TER_PI, HALF_PI, and TWO_PI can be used to replace the radian values for 180°, 45°, 90°, and 360°.

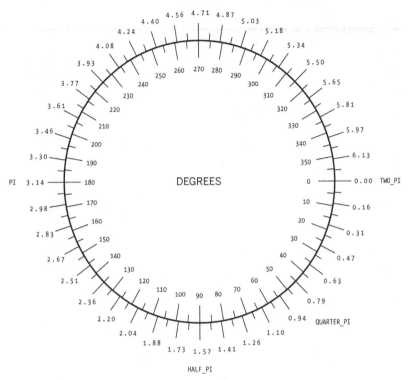

RADIANS

Figure 3-2. *Radians and degrees are two ways to measure an angle. Degrees move around the circle from 0 to 360, while radians measure the angles in relation to pi, from 0 to approximately 6.28.*

Example 3-8: Draw with Degrees

If you prefer to use degree measurements, you can convert to radians with the `radians()` function. This function takes an angle in degrees and changes it to the corresponding radian value. The following example is the same as Example 3-7 on page 18, but it uses the `radians()` function to define the start and stop values in degrees:

```
size(480, 120);
arc(90, 60, 80, 80, 0, radians(90));
```

```
arc(190, 60, 80, 80, 0, radians(270));
arc(290, 60, 80, 80, radians(180), radians(450));
arc(390, 60, 80, 80, radians(45), radians(225));
```

Drawing Order

When a program runs, the computer starts at the top and reads each line of code until it reaches the last line and then stops. If you want a shape to be drawn on top of all other shapes, it needs to follow the others in the code.

Example 3-9: Control Your Drawing Order

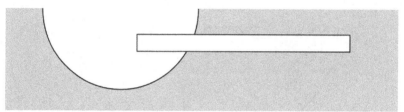

```
size(480, 120);
ellipse(140, 0, 190, 190);
// The rectangle draws on top of the ellipse
// because it comes after in the code
rect(160, 30, 260, 20);
```

Example 3-10: Put It in Reverse

Modify by reversing the order of **rect()** and **ellipse()** to see the circle on top of the rectangle:

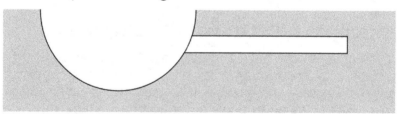

```
size(480, 120);
rect(160, 30, 260, 20);
// The ellipse draws on top of the rectangle
```

```
// because it comes after in the code
ellipse(140, 0, 190, 190);
```

You can think of it like painting with a brush or making a collage. The last element that you add is what's visible on top.

Shape Properties

The most basic and useful shape properties are stroke weight, the way the ends (caps) of lines are drawn, and how the corners of shapes are displayed.

Example 3-11: Set Stroke Weight

The default stroke weight is a single pixel, but this can be changed with the strokeWeight() function. The single parameter to strokeWeight() sets the width of drawn lines:

```
size(480, 120);
ellipse(75, 60, 90, 90);
strokeWeight(8);  // Stroke weight to 8 pixels
ellipse(175, 60, 90, 90);
ellipse(279, 60, 90, 90);
strokeWeight(20);  // Stroke weight to 20 pixels
ellipse(389, 60, 90, 90);
```

Example 3-12: Set Stroke Caps

The strokeCap() function changes how lines are drawn at their endpoints. By default, they have rounded ends:

```
size(480, 120);
strokeWeight(24);
line(60, 25, 130, 95);
strokeCap(SQUARE);    // Square the line endings
line(160, 25, 230, 95);
strokeCap(PROJECT);   // Project the line endings
line(260, 25, 330, 95);
strokeCap(ROUND);     // Round the line endings
line(360, 25, 430, 95);
```

Example 3-13: Set Stroke Joins

The strokeJoin() function changes the way lines are joined (how the corners look). By default, they have pointed (mitered) corners:

```
size(480, 120);
strokeWeight(12);
rect(60, 25, 70, 70);
strokeJoin(ROUND);    // Round the stroke corners
rect(160, 25, 70, 70);
strokeJoin(BEVEL);    // Bevel the stroke corners
rect(260, 25, 70, 70);
strokeJoin(MITER);    // Miter the stroke corners
rect(360, 25, 70, 70);
```

When any of these attributes are set, all shapes drawn afterward are affected. For instance, in Example 3-11 on page 21, notice how the second and third circles both have the same stroke weight, even though the weight is set only once before both are drawn.

Drawing Modes

A group of functions with "mode" in their name change how Processing draws geometry to the screen. In this chapter, we'll look at ellipseMode() and rectMode(), which help us to draw ellipses and rectangles, respectively; later in the book, we'll cover imageMode() and shapeMode().

Example 3-14: On the Corner

By default, the ellipse() function uses its first two parameters as the *x* and *y* coordinate of the center and the third and fourth parameters as the width and height. After ellipseMode(CORNER) is run in a sketch, the first two parameters to ellipse() then define the position of the upper-left corner of the rectangle the ellipse is inscribed within. This makes the ellipse() function behave more like rect() as seen in this example:

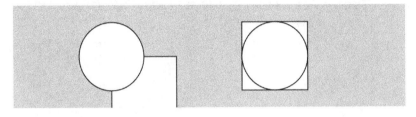

```
size(480, 120);
rect(120, 60, 80, 80);
ellipse(120, 60, 80, 80);
ellipseMode(CORNER);
rect(280, 20, 80, 80);
ellipse(280, 20, 80, 80);
```

You'll find these "mode" functions in examples throughout the book. There are more options for how to use them in the *Processing Reference*.

Color

All the shapes so far have been filled white with black outlines, and the background of the Display Window has been light gray. To change them, use the background(), fill(), and stroke() functions. The values of the parameters are in the range of 0 to 255, where 255 is white, 128 is medium gray, and 0 is black. Figure 3-3 shows how the values from 0 to 255 map to different gray levels.

R	G	B
255	204	0
249	201	4
243	199	9
238	197	13
232	194	18
226	192	22
221	190	27
215	188	31
209	185	36
204	183	40
198	181	45
192	179	49
187	176	54
181	174	58
175	172	63
170	170	68
164	167	72
158	165	77
153	163	81
147	160	86
141	158	90
136	156	95
130	154	99
124	151	104
119	149	108
113	147	113
107	145	117
102	142	122
96	140	126
90	138	131
85	136	136
79	133	140
73	131	145
68	129	149
62	126	154
56	124	158
51	122	163
45	120	167
39	117	172
34	115	176
28	113	181
22	111	185
17	108	190
11	106	194
5	104	199
0	102	204

R	G	B
0	102	204
5	105	205
11	108	206
17	112	207
22	115	208
28	119	209
34	122	210
39	125	211
45	129	213
51	132	214
56	136	215
62	139	216
68	142	217
73	146	218
79	149	219
85	153	221
90	156	222
96	159	223
102	163	224
107	166	225
113	170	226
119	173	227
124	176	228
130	180	230
136	183	231
141	187	232
147	190	233
153	193	234
158	197	235
164	200	236
170	204	238
175	207	239
181	210	240
187	214	241
192	217	242
198	221	243
204	224	244
209	227	245
215	231	247
221	234	248
226	238	249
232	241	250
238	244	251
243	248	252
249	251	253
255	255	255

Figure 3-3. *Colors are created by defining RGB (red, green, blue) values*

Example 3-15: Paint with Grays

This example shows three different gray values on a black background:

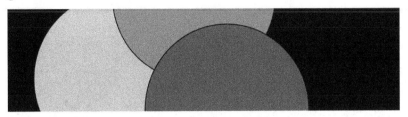

```
size(480, 120);
background(0);               // Black
fill(204);                   // Light gray
ellipse(132, 82, 200, 200); // Light gray circle
fill(153);                   // Medium gray
ellipse(228, -16, 200, 200); // Medium gray circle
fill(102);                   // Dark gray
ellipse(268, 118, 200, 200); // Dark gray circle
```

Example 3-16: Control Fill and Stroke

You can disable the stroke so that there's no outline by using noStroke(), and you can disable the fill of a shape with noFill():

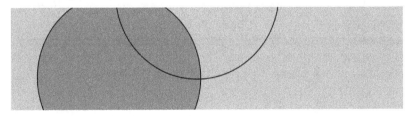

```
size(480, 120);
fill(153);                   // Medium gray
ellipse(132, 82, 200, 200); // Gray circle
noFill();                    // Turn off fill
ellipse(228, -16, 200, 200); // Outline circle
noStroke();                  // Turn off stroke
ellipse(268, 118, 200, 200); // Doesn't draw!
```

Be careful not to disable the fill and stroke at the same time, as we've done in the previous example, because nothing will draw to the screen.

Example 3-17: Draw with Color

To move beyond grayscale values, you use three parameters to specify the red, green, and blue components of a color.

Run the code in Processing to reveal the colors:

```
size(480, 120);
noStroke();
background(0, 26, 51);        // Dark blue color
fill(255, 0, 0);              // Red color
ellipse(132, 82, 200, 200);   // Red circle
fill(0, 255, 0);             // Green color
ellipse(228, -16, 200, 200);  // Green circle
fill(0, 0, 255);             // Blue color
ellipse(268, 118, 200, 200);  // Blue circle
```

This is referred to as RGB color, which comes from how computers define colors on the screen. The three numbers stand for the values of red, green, and blue, and they range from 0 to 255 the way that the gray values do. Using RGB color isn't very intuitive, so to choose colors, use Tools→Color Selector, which shows a color palette similar to those found in other software (see Figure 3-4). Select a color, and then use the R, G, and B values as the parameters for your background(), fill(), or stroke() function.

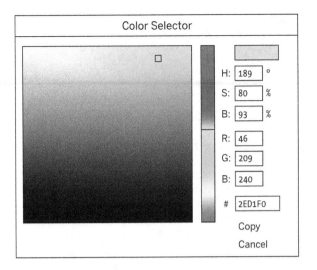

Figure 3-4. *Processing Color Selector*

Example 3-18: Set Transparency

By adding an optional fourth parameter to fill() or stroke(), you can control the transparency. This fourth parameter is known as the *alpha* value, and also uses the range 0 to 255 to set the amount of transparency. The value 0 defines the color as entirely transparent (it won't display), the value 255 is entirely opaque, and the values between these extremes cause the colors to mix on screen:

```
size(480, 120);
noStroke();
background(204, 226, 225);      // Light blue color
fill(255, 0, 0, 160);           // Red color
ellipse(132, 82, 200, 200);     // Red circle
fill(0, 255, 0, 160);           // Green color
ellipse(228, -16, 200, 200);    // Green circle
```

```
fill(0, 0, 255, 160);        // Blue color
ellipse(268, 118, 200, 200); // Blue circle
```

Custom Shapes

You're not limited to using these basic geometric shapes—you can also define new shapes by connecting a series of points.

Example 3-19: Draw an Arrow

The beginShape() function signals the start of a new shape. The vertex() function is used to define each pair of x and y coordinates for the shape. Finally, endShape() is called to signal that the shape is finished:

```
size(480, 120);
beginShape();
fill(153, 176, 180);
vertex(180, 82);
vertex(207, 36);
vertex(214, 63);
vertex(407, 11);
vertex(412, 30);
vertex(219, 82);
vertex(226, 109);
endShape();
```

Example 3-20: Close the Gap

When you run Example 3-19 on page 28, you'll see the first and last point are not connected. To do this, add the word CLOSE as a parameter to endShape(), like this:

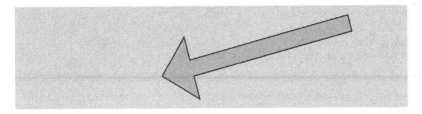

```
size(480, 120);
beginShape();
fill(153, 176, 180);
vertex(180, 82);
vertex(207, 36);
vertex(214, 63);
vertex(407, 11);
vertex(412, 30);
vertex(219, 82);
vertex(226, 109);
endShape(CLOSE);
```

Example 3-21: Create Some Creatures

The power of defining shapes with vertex() is the ability to make shapes with complex outlines. Processing can draw thousands and thousands of lines at a time to fill the screen with fantastic shapes that spring from your imagination. A modest but more complex example follows:

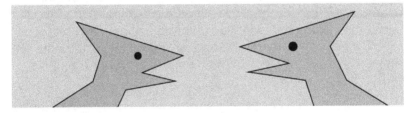

```
size(480, 120);

// Left creature
fill(153, 176, 180);
beginShape();
vertex(50, 120);
vertex(100, 90);
vertex(110, 60);
vertex(80, 20);
vertex(210, 60);
```

```
vertex(160, 80);
vertex(200, 90);
vertex(140, 100);
vertex(130, 120);
endShape();
fill(0);
ellipse(155, 60, 8, 8);

// Right creature
fill(176, 186, 163);
beginShape();
vertex(370, 120);
vertex(360, 90);
vertex(290, 80);
vertex(340, 70);
vertex(280, 50);
vertex(420, 10);
vertex(390, 50);
vertex(410, 90);
vertex(460, 120);
endShape();
fill(0);
ellipse(345, 50, 10, 10);
```

Comments

The examples in this chapter use double slashes (//) at the end of a line to add comments to the code. Comments are parts of the program that are ignored when the program is run. They are useful for making notes for yourself that explain what's happening in the code. If others are reading your code, comments are especially important to help them understand your thought process.

Comments are also especially useful for a number of different options, such as when trying to choose the right color. So, for instance, I might be trying to find just the right red for an ellipse:

```
size(200, 200);
fill(165, 57, 57);
ellipse(100, 100, 80, 80);
```

Now suppose I want to try a different red, but don't want to lose the old one. I can copy and paste the line, make a change, and then "comment out" the old one:

```
size(200, 200);
//fill(165, 57, 57);
fill(144, 39, 39);
ellipse(100, 100, 80, 80);
```

Placing // at the beginning of the line temporarily disables it. Or I can remove the // and place it in front of the other line if I want to try it again:

```
size(200, 200);
fill(165, 57, 57);
//fill(144, 39, 39);
ellipse(100, 100, 80, 80);
```

As you work with Processing sketches, you'll find yourself creating dozens of iterations of ideas; using comments to make notes or to disable code can help you keep track of multiple options.

As a shortcut, you can also use Ctrl-/ (Cmd-/ on the Mac) to add or remove comments from the current line or a selected block of text. You can also comment out many lines at a time with the alternative comment notation introduced in "Comments" on page 203.

Robot 1: Draw

This is P5, the Processing Robot. There are 10 different programs to draw and animate him in the book—each one explores a different programming idea. P5's design was inspired by Sputnik I (1957), Shakey from the Stanford Research Institute (1966–1972), the fighter drone in David Lynch's *Dune* (1984), and HAL 9000 from *2001: A Space Odyssey* (1968), among other robot favorites.

The first robot program uses the drawing functions introduced in this chapter. The parameters to the fill() and stroke() functions set the gray values. The line(), ellipse(), and rect() functions define the shapes that create the robot's neck, antennae, body, and head. To get more familiar with the functions, run the program and change the values to redesign the robot:

```
size(720, 480);
strokeWeight(2);
background(0, 153, 204);      // Blue background
ellipseMode(RADIUS);

// Neck
stroke(255);                  // Set stroke to white
line(266, 257, 266, 162);     // Left
```

```
line(276, 257, 276, 162);      // Middle
line(286, 257, 286, 162);      // Right

// Antennae
line(276, 155, 246, 112);      // Small
line(276, 155, 306, 56);       // Tall
line(276, 155, 342, 170);      // Medium

// Body
noStroke();                    // Disable stroke
fill(255, 204, 0);             // Set fill to orange
ellipse(264, 377, 33, 33);     // Antigravity orb
fill(0);                       // Set fill to black
rect(219, 257, 90, 120);       // Main body
fill(255, 204, 0);             // Set fill to yellow
rect(219, 274, 90, 6);         // Yellow stripe

// Head
fill(0);                       // Set fill to black
ellipse(276, 155, 45, 45);     // Head
fill(255);                     // Set fill to white
ellipse(288, 150, 14, 14);     // Large eye
fill(0);                       // Set fill to black
ellipse(288, 150, 3, 3);       // Pupil
fill(153, 204, 255);           // Set fill to light blue
ellipse(263, 148, 5, 5);       // Small eye 1
ellipse(296, 130, 4, 4);       // Small eye 2
ellipse(305, 162, 3, 3);       // Small eye 3
```

4/Variables

A *variable* stores a value in memory so that it can be used later in a program. The variable can be used many times within a single program, and the value is easily changed while the program is running.

First Variables

One of the reasons we use variables is to avoid repeating ourselves in the code. If you are typing the same number more than once, consider using a variable instead so that your code is more general and easier to update.

Example 4-1: Reuse the Same Values

For instance, when you make the *y* coordinate and diameter for the three circles in this example into variables, the same values are used for each ellipse:

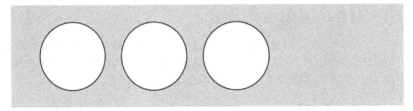

```
size(480, 120);
int y = 60;
int d = 80;
ellipse(75, y, d, d);     // Left
ellipse(175, y, d, d);    // Middle
ellipse(275, y, d, d);    // Right
```

Example 4-2: Change Values

Simply changing the *y* and *d* variables alters all three ellipses:

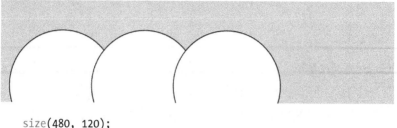

```
size(480, 120);
int y = 100;
int d = 130;
ellipse(75, y, d, d);      // Left
ellipse(175, y, d, d);     // Middle
ellipse(275, y, d, d);     // Right
```

Without the variables, you'd need to change the *y* coordinate used in the code three times and the diameter six times. When comparing Example 4-1 on page 35 and Example 4-2 on page 36, notice how the bottom three lines are the same, and only the middle two lines with the variables are different. Variables allow you to separate the lines of the code that change from the lines that don't, which makes programs easier to modify. For instance, if you place variables that control colors and sizes of shapes in one place, then you can quickly explore different visual options by focusing on only a few lines of code.

Making Variables

When you make your own variables, you determine the *name*, the *data type*, and the *value*. The name is what you decide to call the variable. Choose a name that is informative about what the variable stores, but be consistent and not too verbose. For instance, the variable name "radius" will be clearer than "r" when you look at the code later.

The range of values that can be stored within a variable is defined by its *data type*. For instance, the *integer* data type can store numbers without decimal places (whole numbers). In code, *integer* is abbreviated to int. There are data types to store each kind of data: integers, floating-point (decimal) numbers, characters, words, images, fonts, and so on.

Variables must first be *declared*, which sets aside space in the computer's memory to store the information. When declaring a variable, you also need to specify its data type (such as `int`), which indicates what kind of information is being stored. After the data type and name are set, a value can be assigned to the variable:

```
int x;  // Declare x as an int variable
x = 12; // Assign a value to x
```

This code does the same thing, but is shorter:

```
int x = 12; // Declare x as an int variable and assign a value
```

The name of the data type is included on the line of code that declares a variable, but it's not written again. Each time the data type is written in front of the variable name, the computer thinks you're trying to declare a new variable. You can't have two variables with the same name in the same part of the program (Appendix D), so the program has an error:

```
int x;      // Declare x as an int variable
int x = 12; // ERROR! Can't have two variables called x here
```

Each data type stores a different kind of data. For instance, an `int` variable can store a whole number, but it can't store a number with decimal points, called a `float`. The word "`float`" refers to "floating point," which describes the technique used to store a number with decimal points in memory. (The specifics of that technique aren't important here.)

A floating-point number can't be assigned to an `int` because information would be lost. For instance, the value 12.2 is different from its nearest `int` equivalent, the value 12. In code, this operation will create an error:

```
int x = 12.2; // ERROR! A floating-point value can't fit in
an int
```

However, a `float` variable can store an integer. For instance, the integer value 12 can be converted to the floating-point equivalent 12.0 because no information is lost. This code works without an error:

```
float x = 12;  // Automatically converts 12 to 12.0
```

Data types are discussed in more detail in Appendix B.

Processing Variables

Processing has a series of special variables to store information about the program while it runs. For instance, the width and height of the window are stored in variables called `width` and `height`. These values are set by the `size()` function. They can be used to draw elements relative to the size of the window, even if the `size()` line changes.

Example 4-3: Adjust the Size, See What Follows

In this example, change the parameters to `size()` to see how it works:

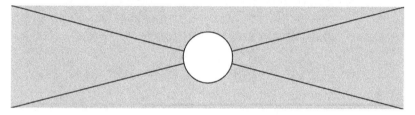

```
size(480, 120);
line(0, 0, width, height); // Line from (0,0) to (480, 120)
line(width, 0, 0, height); // Line from (480, 0) to (0, 120)
ellipse(width/2, height/2, 60, 60);
```

Other special variables keep track of the status of the mouse and keyboard values and much more. These are discussed in Chapter 5.

A Little Math

People often assume that math and programming are the same thing. Although knowledge of math can be useful for certain types of coding, basic arithmetic covers the most important parts.

Example 4-4: Basic Arithmetic

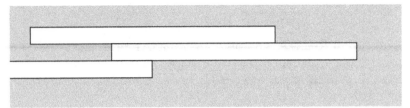

```
size(480, 120);
int x = 25;
int h = 20;
int y = 25;
rect(x, y, 300, h);          // Top
x = x + 100;
rect(x, y + h, 300, h);      // Middle
x = x - 250;
rect(x, y + h*2, 300, h);    // Bottom
```

In code, symbols like +, −, and * are called *operators*. When placed between two values, they create an *expression*. For instance, 5 + 9 and 1024 − 512 are both expressions. The operators for the basic math operations are:

+	Addition
−	Subtraction
*	Multiplication
/	Division
=	Assignment

Processing has a set of rules to define which operators take precedence over others, meaning which calculations are made first, second, third, and so on. These rules define the order in which the code is run. A little knowledge about this goes a long way toward understanding how a short line of code like this works:

```
int x = 4 + 4 * 5; // Assign 24 to x
```

The expression 4 * 5 is evaluated first because multiplication has the highest priority. Second, 4 is added to the product of 4 * 5 to yield 24. Last, because the *assignment operator* (the equals sign) has the lowest precedence, the value 24 is assigned to the variable *x*. This is clarified with parentheses, but the result is the same:

```
int x = 4 + (4 * 5); // Assign 24 to x
```

If you want to force the addition to happen first, just move the parentheses. Because parentheses have a higher precedence than multiplication, the order is changed and the calculation is affected:

```
int x = (4 + 4) * 5; // Assign 40 to x
```

An acronym for this order is often taught in math class: PEMDAS, which stands for Parentheses, Exponents, Multiplication, Division, Addition, Subtraction, where parentheses have the highest priority and subtraction the lowest. The complete order of operations is found in Appendix C.

Some calculations are used so frequently in programming that shortcuts have been developed; it's always nice to save a few keystrokes. For instance, you can add to a variable, or subtract from it, with a single operator:

```
x += 10; // This is the same as x = x + 10
y -= 15; // This is the same as y = y - 15
```

It's also common to add or subtract 1 from a variable, so shortcuts exist for this as well. The ++ and -- operators do this:

```
x++; // This is the same as x = x + 1
y--; // This is the same as y = y - 1
```

More shortcuts can be found in the *Processing Reference*.

Repetition

As you write more programs, you'll notice that patterns occur when lines of code are repeated, but with slight variations. A code structure called a *for loop* makes it possible to run a line of code more than once to condense this type of repetition into fewer lines. This makes your programs more modular and easier to change.

Example 4-5: Do the Same Thing Over and Over

This example has the type of pattern that can be simplified with a for loop:

```
size(480, 120);
strokeWeight(8);
line(20, 40, 80, 80);
line(80, 40, 140, 80);
line(140, 40, 200, 80);
line(200, 40, 260, 80);
line(260, 40, 320, 80);
line(320, 40, 380, 80);
line(380, 40, 440, 80);
```

Example 4-6: Use a for Loop

The same thing can be done with a for loop, and with less code:

```
size(480, 120);
strokeWeight(8);
for (int i = 20; i < 400; i += 60) {
  line(i, 40, i + 60, 80);
}
```

The for loop is different in many ways from the code we've written so far. Notice the braces, the { and } characters. The code between the braces is called a *block*. This is the code that will be repeated on each iteration of the for loop.

Inside the parentheses are three statements, separated by semicolons, that work together to control how many times the code inside the block is run. From left to right, these statements are referred to as the *initialization* (init), the *test*, and the *update*:

```
for (init; test; update) {
    statements
}
```

The init sets the starting value, often declaring a new variable to use within the for loop. In the earlier example, an integer named i was declared and set to 20. The variable name i is frequently used, but there's really nothing special about it. The *test* evaluates the value of this variable (here, it checks whether i still less than 400), and the *update* changes the variable's value (adding 60 before repeating the loop). Figure 4-1 shows the order in which they run and how they control the code statements inside the block.

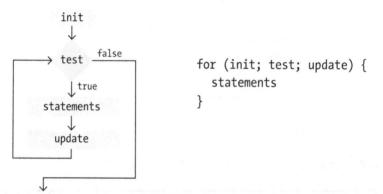

```
for (init; test; update) {
    statements
}
```

Figure 4-1. *Flow diagram of a for loop*

The *test* statement requires more explanation. It's always a *relational expression* that compares two values with a *relational operator*. In this example, the expression is "i < 400" and the operator is the < (less than) symbol. The most common relational operators are:

>	Greater than
<	Less than
>=	Greater than or equal to
<=	Less than or equal to
==	Equal to
!=	Not equal to

The relational expression always evaluates to true or false. For instance, the expression 5 > 3 is true. We can ask the question, "Is five greater than three?" Because the answer is "yes," we say the expression is true. For the expression 5 < 3, we ask, "Is five less than three?" Because the answer is "no," we say the expression is false. When the evaluation is true, the code inside the block is run, and when it's false, the code inside the block is not run and the for loop ends.

Example 4-7: Flex Your for Loop's Muscles

The ultimate power of working with a for loop is the ability to make quick changes to the code. Because the code inside the block is typically run multiple times, a change to the block is magnified when the code is run. By modifying Example 4-6 on page 41 only slightly, we can create a range of different patterns:

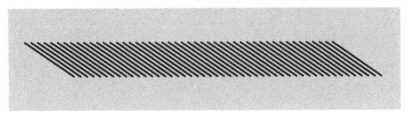

```
size(480, 120);
strokeWeight(2);
for (int i = 20; i < 400; i += 8) {
  line(i, 40, i + 60, 80);
}
```

Example 4-8: Fanning Out the Lines

```
size(480, 120);
strokeWeight(2);
```

```
for (int i = 20; i < 400; i += 20) {
  line(i, 0, i + i/2, 80);
}
```

Example 4-9: Kinking the Lines

```
size(480, 120);
strokeWeight(2);
for (int i = 20; i < 400; i += 20) {
  line(i, 0, i + i/2, 80);
  line(i + i/2, 80, i*1.2, 120);
}
```

Example 4-10: Embed One for Loop in Another

When one **for** loop is embedded inside another, the number of repetitions is multiplied. First, let's look at a short example, and then we'll break it down in Example 4-11 on page 45:

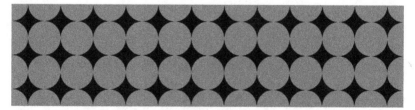

```
size(480, 120);
background(0);
noStroke();
for (int y = 0; y <= height; y += 40) {
  for (int x = 0; x <= width; x += 40) {
    fill(255, 140);
    ellipse(x, y, 40, 40);
```

```
  }
}
```

Example 4-11: Rows and Columns

In this example, the for loops are adjacent, rather than one embedded inside the other. The result shows that one for loop is drawing a column of 4 circles and the other is drawing a row of 13 circles:

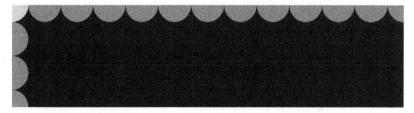

```
size(480, 120);
background(0);
noStroke();
for (int y = 0; y < height+45; y += 40) {
  fill(255, 140);
  ellipse(0, y, 40, 40);
}
for (int x = 0; x < width+45; x += 40) {
  fill(255, 140);
  ellipse(x, 0, 40, 40);
}
```

When one of these for loops is placed inside the other, as in Example 4-10 on page 44, the 4 repetitions of the first loop are compounded with the 13 of the second in order to run the code inside the embedded block 52 times (4×13 = 52).

Example 4-10 on page 44 is a good base for exploring many types of repeating visual patterns. The following examples show a couple of ways that it can be extended, but this is only a tiny sample of what's possible. In Example 4-12 on page 46, the code draws a line from each point in the grid to the center of the screen. In Example 4-13 on page 46, the ellipses shrink with each new row and are moved to the right by adding the y coordinate to the x coordinate.

Example 4-12: Pins and Lines

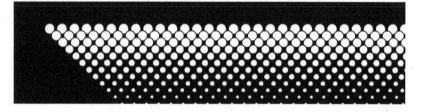

```
size(480, 120);
background(0);
fill(255);
stroke(102);
for (int y = 20; y <= height-20; y += 10) {
  for (int x = 20; x <= width-20; x += 10) {
    ellipse(x, y, 4, 4);
    // Draw a line to the center of the display
    line(x, y, 240, 60);
  }
}
```

Example 4-13: Halftone Dots

```
size(480, 120);
background(0);
for (int y = 32; y <= height; y += 8) {
  for (int x = 12; x <= width; x += 15) {
    ellipse(x + y, y, 16 - y/10.0, 16 - y/10.0);
  }
}
```

Robot 2: Variables

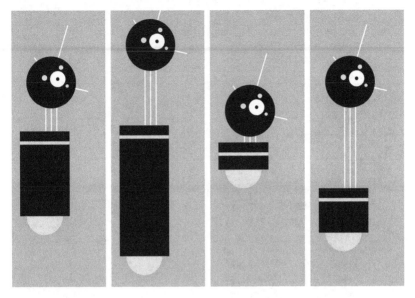

The variables introduced in this program make the code look more difficult than Robot 1 (see "Robot 1: Draw" on page 32), but now it's much easier to modify, because numbers that depend on one another are in a single location. For instance, the neck can be drawn based on the bodyHeight variable. The group of variables at the top of the code control the aspects of the robot that we want to change: location, body height, and neck height. You can see some of the range of possible variations in the figure; from left to right, here are the values that correspond to them:

y = 390	y = 460	y = 310	y = 420
bodyHeight = 180	bodyHeight = 260	bodyHeight = 80	bodyHeight = 110
neckHeight = 40	neckHeight = 95	neckHeight = 10	neckHeight = 140

When altering your own code to use variables instead of numbers, plan the changes carefully, then make the modifications in short steps. For instance, when this program was written, each variable was created one at a time to minimize the complexity of the transition. After a variable was added and the code was run to ensure it was working, the next variable was added:

```
int x = 60;            // x coordinate
int y = 390;           // y coordinate
int bodyHeight = 180;  // Body height
int neckHeight = 40;   // Neck height
int radius = 45;
int ny = y - bodyHeight - neckHeight - radius;  // Neck y

size(170, 480);
strokeWeight(2);
background(0, 153, 204);
ellipseMode(RADIUS);

// Neck
stroke(255);
line(x+2, y-bodyHeight, x+2, ny);
line(x+12, y-bodyHeight, x+12, ny);
line(x+22, y-bodyHeight, x+22, ny);

// Antennae
line(x+12, ny, x-18, ny-43);
line(x+12, ny, x+42, ny-99);
line(x+12, ny, x+78, ny+15);

// Body
noStroke();
fill(255, 204, 0);
ellipse(x, y-33, 33, 33);
fill(0);
rect(x-45, y-bodyHeight, 90, bodyHeight-33);
fill(255, 204, 0);
rect(x-45, y-bodyHeight+17, 90, 6);

// Head
fill(0);
ellipse(x+12, ny, radius, radius);
fill(255);
ellipse(x+24, ny-6, 14, 14);
fill(0);
ellipse(x+24, ny-6, 3, 3);
fill(153, 204, 255);
ellipse(x, ny-8, 5, 5);
ellipse(x+30, ny-26, 4, 4);
ellipse(x+41, ny+6, 3, 3);
```

5/Response

Code that responds to input from the mouse, keyboard, and other devices has to run continuously. To make this happen, place the lines that update inside a Processing function called **draw()**.

Once and Forever

The code within the **draw()** block runs from top to bottom, then repeats until you quit the program by clicking the Stop button or closing the window. Each trip through **draw()** is called a *frame*. (The default frame rate is 60 frames per second, but this can be changed).

Example 5-1: The draw() Function

To see how **draw()** works, run this example:

```
void draw() {
  // Displays the frame count to the Console
  println("I'm drawing");
  println(frameCount);
}
```

You'll see the following:

```
I'm drawing
1
I'm drawing
2
I'm drawing
3
...
```

In the preceding example program, the **println()** functions write the text "I'm drawing" followed by the current frame count as counted by the special `frameCount` variable (1, 2, 3, ...). The text appears in the Console, the black area at the bottom of the Processing editor window.

Example 5-2: The setup() Function

To complement the looping `draw()` function, Processing has a function called `setup()` that runs just once when the program starts:

```
void setup() {
  println("I'm starting");
}

void draw() {
  println("I'm running");
}
```

When this code is run, the following is written to the Console:

```
I'm starting
I'm running
I'm running
I'm running
...
```

The text "I'm running" continues to write to the Console until the program is stopped.

In a typical program, the code inside `setup()` is used to define the starting values. The first line is always the `size()` function, often followed by code to set the starting fill and stroke colors, or perhaps to load images and fonts. (If you don't include the `size()` function, the Display Window will be 100×100 pixels.)

Now you know how to use `setup()` and `draw()`, but this isn't the whole story. There's one more location to put code—you can also place variables outside of `setup()` and `draw()`. If you create a variable inside of `setup()`, you can't use it inside of `draw()`, so you need to place those variables somewhere else. Such variables are called *global* variables, because they can be used anywhere ("globally") in the program. This is clearer when we list the order in which the code is run:

1. Variables declared outside of `setup()` and `draw()` are created.
2. Code inside `setup()` is run once.
3. Code inside `draw()` is run continuously.

Example 5-3: Global Variable

The following example puts it all together:

```
int x = 280;
int y = -100;
int diameter = 380;

void setup() {
  size(480, 120);
  fill(102);
}

void draw() {
  background(204);
  ellipse(x, y, diameter, diameter);
}
```

Follow

Now that we have code running continuously, we can track the mouse position and use those numbers to move elements on screen.

Example 5-4: Track the Mouse

The `mouseX` variable stores the *x* coordinate, and the `mouseY` variable stores the *y* coordinate:

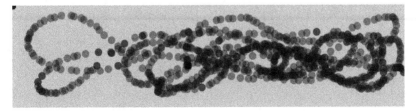

```
void setup() {
  size(480, 120);
  fill(0, 102);
  noStroke();
}

void draw() {
  ellipse(mouseX, mouseY, 9, 9);
}
```

In this example, each time the code in the draw() block is run, a new circle is drawn to the window. This image was made by moving the mouse around to control the circle's location. Because the fill is set to be partially transparent, denser black areas show where the mouse spent more time and where it moved slowly. The circles that are spaced farther apart show when the mouse was moving faster.

Example 5-5: The Dot Follows You

In this example, a new circle is added to the window each time the code in draw() is run. To refresh the screen and only display the newest circle, place a background() function at the beginning of draw() before the shape is drawn:

```
void setup() {
  size(480, 120);
  fill(0, 102);
  noStroke();
}

void draw() {
  background(204);
  ellipse(mouseX, mouseY, 9, 9);
}
```

The `background()` function clears the entire window, so be sure to always place it before other functions inside `draw()`; otherwise, the shapes drawn before it will be erased.

Example 5-6: Draw Continuously

The `pmouseX` and `pmouseY` variables store the position of the mouse at the previous frame. Like `mouseX` and `mouseY`, these special variables are updated each time `draw()` runs. When combined, they can be used to draw continuous lines by connecting the current and most recent location:

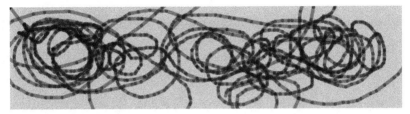

```
void setup() {
  size(480, 120);
  strokeWeight(4);
  stroke(0, 102);
}

void draw() {
  line(mouseX, mouseY, pmouseX, pmouseY);
}
```

Example 5-7: Set Line Thickness

The `pmouseX` and `pmouseY` variables can also be used to calculate the speed of the mouse. This is done by measuring the distance between the current and most recent mouse location. If the mouse is moving slowly, the distance is small, but if the mouse starts moving faster, the distance grows. A function called `dist()` simplifies this calculation, as shown in the following example. Here, the speed of the mouse is used to set the thickness of the drawn line:

```
void setup() {
  size(480, 120);
  stroke(0, 102);
}

void draw() {
  float weight = dist(mouseX, mouseY, pmouseX, pmouseY);
  strokeWeight(weight);
  line(mouseX, mouseY, pmouseX, pmouseY);
}
```

Example 5-8: Easing Does It

In Example 5-7 on page 53, the values from the mouse are converted directly into positions on the screen. But sometimes you want the values to follow the mouse loosely—to lag behind to create a more fluid motion. This technique is called *easing*. With easing, there are two values: the current value and the value to move toward (see Figure 5-1). At each step in the program, the current value moves a little closer to the target value:

```
float x;
float easing = 0.01;

void setup() {
  size(220, 120);
}

void draw() {
  float targetX = mouseX;
  x += (targetX - x) * easing;
  ellipse(x, 40, 12, 12);
  println(targetX + " : " + x);
}
```

The value of the x variable is always getting closer to targetX. The speed at which it catches up with targetX is set with the eas ing variable, a number between 0 and 1. A small value for easing causes more of a delay than a larger value. With an easing value of 1, there is no delay. When you run Example 5-8 on page 54, the actual values are shown in the Console through the println() function. When moving the mouse, notice how the numbers are far apart, but when the mouse stops moving, the x value gets closer to targetX.

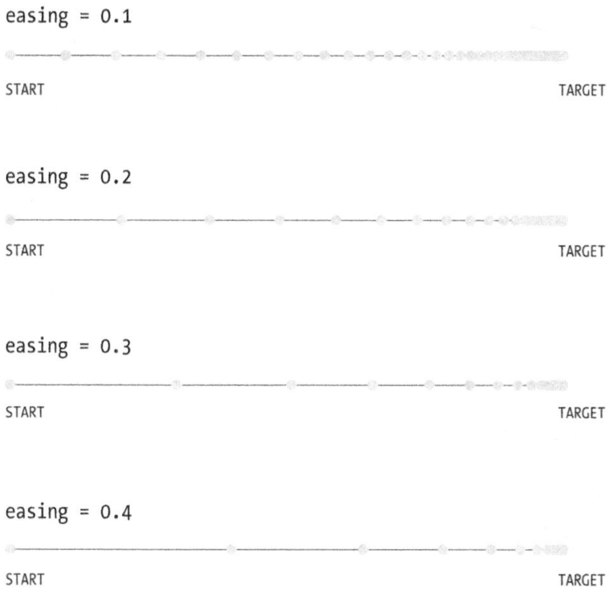

easing = 0.1

START   TARGET

easing = 0.2

START   TARGET

easing = 0.3

START   TARGET

easing = 0.4

START   TARGET

Figure 5-1. *Easing changes the number of steps it takes to move from one place to another*

All of the work in this example happens on the line that begins x +=. There, the difference between the target and current value is calculated, then multiplied by the easing variable and added to x to bring it closer to the target.

Example 5-9: Smooth Lines with Easing

In this example, the easing technique is applied to Example 5-7 on page 53. In comparison, the lines are more fluid:

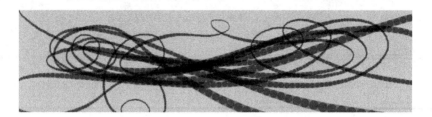

```
float x;
float y;
float px;
float py;
float easing = 0.05;

void setup() {
  size(480, 120);
  stroke(0, 102);
}

void draw() {
  float targetX = mouseX;
  x += (targetX - x) * easing;
  float targetY = mouseY;
  y += (targetY - y) * easing;
  float weight = dist(x, y, px, py);
  strokeWeight(weight);
  line(x, y, px, py);
  py = y;
  px = x;
}
```

Click

In addition to the location of the mouse, Processing also keeps track of whether the mouse button is pressed. The mousePressed variable has a different value when the mouse button is pressed and when it is not. The mousePressed variable is a data type called boolean, which means that it has only two possible values: true and false. The value of mousePressed is true when a button is pressed.

Example 5-10: Click the Mouse

The `mousePressed` variable is used along with the `if` state determine when a line of code will run and when it won't. Try this example before we explain further:

```
void setup() {
  size(240, 120);
  strokeWeight(30);
}

void draw() {
  background(204);
  stroke(102);
  line(40, 0, 70, height);
  if (mousePressed == true) {
    stroke(0);
  }
  line(0, 70, width, 50);
}
```

In this program, the code inside the `if` block runs only when a mouse button is pressed. When a button is not pressed, this code is ignored. Like the `for` loop discussed in "Repetition" on page 40, the `if` also has a *test* that is evaluated to `true` or `false`:

```
if (test) {
  statements
}
```

When the test is `true`, the code inside the block is run; when the test is `false`, the code inside the block is not run. The computer determines whether the test is `true` or `false` by evaluating the expression inside the parentheses. (If you'd like to refresh your memory, the discussion of relational expressions is in Example 4-6 on page 41.)

The == symbol compares the values on the left and right to test whether they are equivalent. This == symbol is different from the assignment operator, the single = symbol. The == symbol asks, "Are these things equal?" and the = symbol sets the value of a variable.

 It's a common mistake, even for experienced programmers, to write = in your code when you mean to write ==. The Processing software won't always warn you when you do this, so be careful.

Alternatively, the test in draw() in can be written like this:

```
if (mousePressed) {
```

Boolean variables, including mousePressed, don't need the explicit comparison with the == operator, because they can be only true or false.

Example 5-11: Detect When Not Clicked

A single if block gives you the choice of running some code or skipping it. You can extend an if block with an else block, allowing your program to choose between two options. The code inside the else block runs when the value of the if block test is false. For instance, the stroke color for a program can be white when the mouse button is not pressed, and can change to black when the button is pressed:

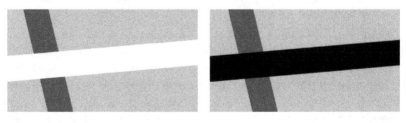

```
void setup() {
  size(240, 120);
  strokeWeight(30);
}

void draw() {
  background(204);
  stroke(102);
  line(40, 0, 70, height);
  if (mousePressed) {
    stroke(0);
  } else {
    stroke(255);
  }
  line(0, 70, width, 50);
}
```

Example 5-12: Multiple Mouse Buttons

Processing also tracks which button is pressed if you have more than one button on your mouse. The mouseButton variable can be one of three values: LEFT, CENTER, or RIGHT. To test which button was pressed, the == operator is needed, as shown here:

```
void setup() {
  size(120, 120);
  strokeWeight(30);
}

void draw() {
  background(204);
  stroke(102);
  line(40, 0, 70, height);
  if (mousePressed) {
    if (mouseButton == LEFT) {
      stroke(255);
```

```
    } else {
      stroke(0);
    }
    line(0, 70, width, 50);
  }
}
```

A program can have many more if and else structures (see
Figure 5-2) than those found in these short examples. They can
be chained together into a long series with each testing for
something different, and if blocks can be embedded inside of
other if blocks to make more complex decisions.

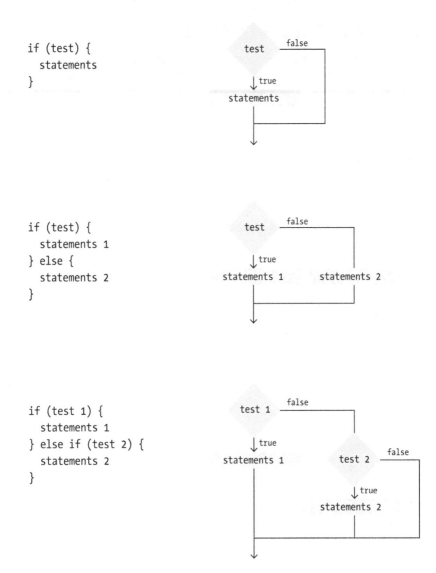

```
if (test) {
   statements
}
```

```
if (test) {
   statements 1
} else {
   statements 2
}
```

```
if (test 1) {
   statements 1
} else if (test 2) {
   statements 2
}
```

Figure 5-2. *The if and else structure makes decisions about which blocks of code to run*

Location

An if structure can be used with the mouseX and mouseY values to determine the location of the cursor within the window.

Example 5-13: Find the Cursor

For instance, this example tests to see whether the cursor is on the left or right side of a line and then moves the line toward the cursor:

```
float x;
int offset = 10;

void setup() {
  size(240, 120);
  x = width/2;
}

void draw() {
  background(204);
  if (mouseX > x) {
    x += 0.5;
    offset = -10;
  }
  if (mouseX < x) {
    x -= 0.5;
    offset = 10;
  }
  // Draw arrow left or right depending on "offset" value
  line(x, 0, x, height);
  line(mouseX, mouseY, mouseX + offset, mouseY - 10);
  line(mouseX, mouseY, mouseX + offset, mouseY + 10);
  line(mouseX, mouseY, mouseX + offset*3, mouseY);
}
```

To write programs that have graphical user interfaces (buttons, checkboxes, scrollbars, etc.), we need to write code that knows when the cursor is within an enclosed area of the screen. The following two examples introduce how to check whether the cursor is inside a circle and a rectangle. The code is written in a modular way with variables, so it can be used to check for *any* circle and rectangle by changing the values.

Example 5-14: The Bounds of a Circle

For the circle test, we use the `dist()` function to get the distance from the center of the circle to the cursor, then we test to see if that distance is less than the radius of the circle (see Figure 5-3). If it is, we know we're inside. In this example, when the cursor is within the area of the circle, its size increases:

```
int x = 120;
int y = 60;
int radius = 12;

void setup() {
  size(240, 120);
  ellipseMode(RADIUS);
}

void draw() {
  background(204);
  float d = dist(mouseX, mouseY, x, y);
  if (d < radius) {
    radius++;
    fill(0);
  } else {
    fill(255);
  }
  ellipse(x, y, radius, radius);
}
```

`dist(x, y, mouseX, mouseY) < radius`

Figure 5-3. *Circle rollover test. When the distance between the mouse and the circle is less than the radius, the mouse is inside the circle.*

Example 5-15: The Bounds of a Rectangle

We use another approach to test whether the cursor is inside a rectangle. We make four separate tests to check if the cursor is on the correct side of each edge of the rectangle, then we compare each test and if they are all **true**, we know the cursor is inside. This is illustrated in Figure 5-4. Each step is simple, but it looks complicated when it's all put together:

```
int x = 80;
int y = 30;
```

```
int w = 80;
int h = 60;

void setup() {
  size(240, 120);
}

void draw() {
  background(204);
  if ((mouseX > x) && (mouseX < x+w) &&
      (mouseY > y) && (mouseY < y+h)) {
    fill(0);
  } else {
    fill(255);
  }
  rect(x, y, w, h);
}
```

The test in the if statement is a little more complicated than we've seen. Four individual tests (e.g., mouseX > x) are combined with the logical AND operator, the && symbol, to ensure that every relational expression in the sequence is true. If one of them is false, the entire test is false and the fill color won't be set to black. This is explained further in the reference entry for &&.

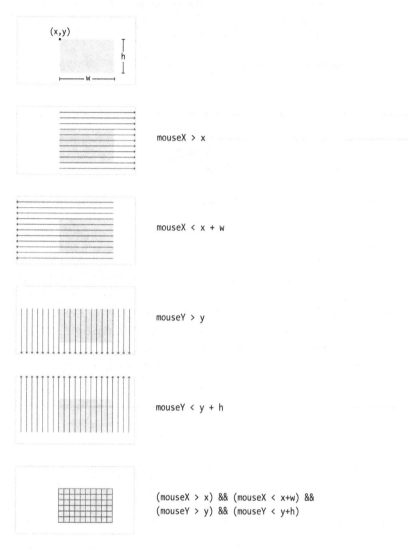

Figure 5-4. *Rectangle rollover test. When all four tests are combined and true, the cursor is inside the rectangle.*

Type

Processing keeps track of when any key on a keyboard is pressed, as well as the last key pressed. Like the `mousePressed` variable, the `keyPressed` variable is `true` when any key is pressed, and `false` when no keys are pressed.

Example 5-16: Tap a Key

In this example, the second line is drawn only when a key is pressed:

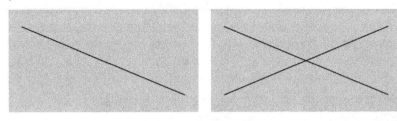

```
void setup() {
  size(240, 120);
}

void draw() {
  background(204);
  line(20, 20, 220, 100);
  if (keyPressed) {
    line(220, 20, 20, 100);
  }
}
```

The key variable stores the most recent key that has been pressed. The data type for key is char, which is short for "character" but usually pronounced like the first syllable of "charcoal." A char variable can store any single character, which includes letters of the alphabet, numbers, and symbols. Unlike a string value (see Example 7-8 on page 97), which is distinguished by double quotes, the char data type is specified by single quotes. This is how a char variable is declared and assigned:

```
char c = 'A';   // Declares and assigns 'A' to the variable c
```

And these attempts will cause an error:

```
char c = "A";   // Error! Can't assign a String to a char
char h = A;     // Error! Missing the single quotes from 'A'
```

Unlike the boolean variable keyPressed, which reverts to false each time a key is released, the key variable keeps its value until the next key is pressed. The following example uses the value of key to draw the character to the screen. Each time a new key is pressed, the value updates and a new character draws. Some keys, like Shift and Alt, don't have a visible character, so when you press them, nothing is drawn.

Example 5-17: Draw Some Letters

This example introduces the textSize() function to set the size of the letters, the textAlign() function to center the text on its *x* coordinate, and the text() function to draw the letter. These functions are discussed in more detail in "Fonts" on page 94:

```
void setup() {
  size(120, 120);
  textSize(64);
  textAlign(CENTER);
}

void draw() {
  background(0);
  text(key, 60, 80);
}
```

By using an if structure, we can test to see whether a specific key is pressed and choose to draw something on screen in response.

Example 5-18: Check for Specific Keys

In this example, we test for an H or N to be typed. We use the comparison operator, the == symbol, to see if the key value is equal to the characters we're looking for:

```
void setup() {
  size(120, 120);
}

void draw() {
  background(204);
  if (keyPressed) {
    if ((key == 'h') || (key == 'H')) {
      line(30, 60, 90, 60);
    }
    if ((key == 'n') || (key == 'N')) {
      line(30, 20, 90, 100);
    }
  }
  line(30, 20, 30, 100);
  line(90, 20, 90, 100);
}
```

When we watch for H or N to be pressed, we need to check for both the lowercase and uppercase letters in the event that someone hits the Shift key or has the Caps Lock set. We combine the two tests together with a logical OR, the || symbol. If we translate the second if statement in this example into plain language, it says, "If the 'h' key is pressed OR the 'H' key is pressed." Unlike with the logical AND (the && symbol), only one of these expressions need be true for the entire test to be true.

Some keys are more difficult to detect, because they aren't tied to a particular letter. Keys like Shift, Alt, and the arrow keys are coded and require an extra step to figure out if they are pressed. First, we need to check if the key that's been pressed is a coded key, then we check the code with the keyCode variable to see which key it is. The most frequently used keyCode values are ALT, CONTROL, and SHIFT, as well as the arrow keys, UP, DOWN, LEFT, and RIGHT.

Example 5-19: Move with Arrow Keys

The following example shows how to check for the left or right arrow keys to move a rectangle:

```
int x = 215;

void setup() {
  size(480, 120);
}

void draw() {
  if (keyPressed && (key == CODED)) {  // If it's a coded key
    if (keyCode == LEFT) {  // If it's the left arrow
      x--;
    } else if (keyCode == RIGHT) {  // If it's the right arrow
      x++;
    }
  }
  rect(x, 45, 50, 50);
}
```

Map

The numbers that are created by the mouse and keyboard often need to be modified to be useful within a program. For instance, if a sketch is 1920 pixels wide and the mouseX values are used to set the color of the background, the range of 0 to 1920 for mouseX might need to move into a range of 0 to 255 to better control the color. This transformation can be done with an equation or with a function called map().

Example 5-20: Map Values to a Range

In this example, the location of two lines are controlled with the mouseX variable. The gray line is synchronized to the cursor position, but the black line stays closer to the center of the screen to move further away from the white line at the left and right edges:

```
void setup() {
  size(240, 120);
  strokeWeight(12);
}

void draw() {
  background(204);
  stroke(102);
  line(mouseX, 0, mouseX, height);  // Gray line
  stroke(0);
  float mx = mouseX/2 + 60;
  line(mx, 0, mx, height);  // Black line
}
```

The map() function is a more general way to make this type of change. It converts a variable from one range of numbers to another. The first parameter is the variable to be converted, the second and third parameters are the low and high values of that variable, and the fourth and fifth parameters are the desired low and high values. The map() function hides the math behind the conversion.

Example 5-21: Map with the map() Function

This example rewrites Example 5-20 on page 70 using map():

```
void setup() {
  size(240, 120);
  strokeWeight(12);
}

void draw() {
  background(204);
  stroke(102);
  line(mouseX, 0, mouseX, height);  // Gray line
  stroke(0);
```

```
float mx = map(mouseX, 0, width, 60, 180);
line(mx, 0, mx, height);  // Black line
}
```

The `map()` function makes the code easy to read, because the minimum and maximum values are clearly written as the parameters. In this example, `mouseX` values between 0 and `width` are converted to a number from 60 (when `mouseX` is 0) up to 180 (when `mouseX` is `width`). You'll find the useful `map()` function in many examples throughout this book.

Robot 3: Response

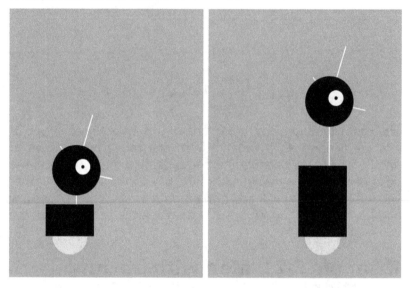

This program uses the variables introduced in Robot 2 (see "Robot 2: Variables" on page 47) and makes it possible to change them while the program runs so that the shapes respond to the mouse. The code inside the `draw()` block runs many times each second. At each frame, the variables defined in the program change in response to the `mouseX` and `mouse Pressed` variables.

The `mouseX` value controls the position of the robot with an easing technique so that movements are less instantaneous and feel more natural. When a mouse button is pressed, the values of `neckHeight` and `bodyHeight` change to make the robot short:

```
float x = 60;          // x coordinate
float y = 440;         // y coordinate
int radius = 45;       // Head radius
int bodyHeight = 160;  // Body height
int neckHeight = 70;   // Neck height

float easing = 0.04;

void setup() {
  size(360, 480);
  ellipseMode(RADIUS);
}

void draw() {
  strokeWeight(2);

  int targetX = mouseX;
  x += (targetX - x) * easing;

  if (mousePressed) {
    neckHeight = 16;
    bodyHeight = 90;
  } else {
    neckHeight = 70;
    bodyHeight = 160;
  }

  float neckY = y - bodyHeight - neckHeight - radius;

  background(0, 153, 204);

  // Neck
  stroke(255);
  line(x+12, y-bodyHeight, x+12, neckY);

  // Antennae
  line(x+12, neckY, x-18, neckY-43);
  line(x+12, neckY, x+42, neckY-99);
  line(x+12, neckY, x+78, neckY+15);

  // Body
  noStroke();
  fill(255, 204, 0);
  ellipse(x, y-33, 33, 33);
  fill(0);
  rect(x-45, y-bodyHeight, 90, bodyHeight-33);
```

```
  // Head
  fill(0);
  ellipse(x+12, neckY, radius, radius);
  fill(255);
  ellipse(x+24, neckY-6, 14, 14);
  fill(0);
  ellipse(x+24, neckY-6, 3, 3);
}
```

6/Translate, Rotate, Scale

Another technique for positioning and moving things on screen is to change the screen coordinate system. For example, you can move a shape 50 pixels to the right, or you can move the location of coordinate (0,0) 50 pixels to the right—the visual result is the same.

```
translate(40, 20);          translate(60, 70);
rect(20, 20, 20, 40);       rect(20, 20, 20, 40);
```

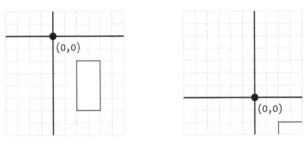

Figure 6-1. *Translating the coordinates*

By modifying the default coordinate system, we can create different *transformations* including *translation*, *rotation*, and *scaling*.

Translate

Working with transformations can be tricky, but the `translate()` function is the most straightforward, so we'll start with that. As

Figure 6-1 shows, this function can shift the coordinate system left, right, up, and down.

Example 6-1: Translating Location

In this example, notice that the rectangle is drawn at coordinate (0,0), but it is moved around on the screen, because it is affected by translate():

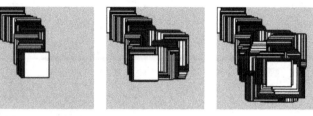

```
void setup() {
  size(120, 120);
}

void draw() {
  translate(mouseX, mouseY);
  rect(0, 0, 30, 30);
}
```

The translate() function sets the (0,0) coordinate of the screen to the mouse location (mouseX and mouseY). Each time the draw() block repeats, the rect() is drawn at the new origin, derived from the current mouse location.

Example 6-2: Multiple Translations

After a transformation is made, it is applied to all drawing functions that follow. Notice what happens when a second translate function is added to control a second rectangle:

```
void setup() {
  size(120, 120);
}

void draw() {
  translate(mouseX, mouseY);
  rect(0, 0, 30, 30);
  translate(35, 10);
  rect(0, 0, 15, 15);
}
```

The values for the **translate()** functions are added together. The smaller rectangle was translated the amount of **mouseX** + 35 and **mouseY** + 10. The x and y coordinates for both rectangles are (0,0), but the **translate()** functions move them to other positions on screen.

However, even though the transformations accumulate within the **draw()** block, they are reset each time **draw()** starts again at the top.

Rotate

The **rotate()** function rotates the coordinate system. It has one parameter, which is the angle (in radians) to rotate. It always rotates relative to (0,0), known as rotating around the origin. Figure 3-2 in Example 3-7 on page 18 shows the radians angle values. Figure 6-2 shows the difference between rotating with positive and negative numbers.

```
rotate(PI/12.0)              rotate(-PI/3);
rect(20, 20, 20, 40);        rect(20, 20, 20, 40);
```

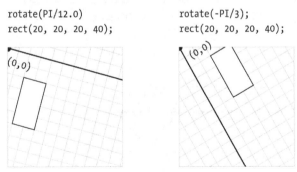

Figure 6-2. *Rotating the coordinates*

Example 6-3: Corner Rotation

To rotate a shape, first define the rotation angle with `rotate()`, then draw the shape. In this sketch, the amount to rotate (`mouseX / 100.0`) will be between 0 and 1.2 to define the rotation angle because `mouseX` will be between 0 and 120, the width of the Display Window specified with the `size()` function. Note that you should divide by 100.0 not 100, because of how numbers work in Processing (see "Making Variables" on page 36).

```
void setup() {
  size(120, 120);
}

void draw() {
  rotate(mouseX / 100.0);
  rect(40, 30, 160, 20);
}
```

Example 6-4: Center Rotation

To rotate a shape around its own center, it must be drawn with coordinate (0,0) in the middle. In this example, because the shape is 160 wide and 20 high as defined in `rect()`, it is drawn at the coordinate (−80, −10) to place (0,0) at the center of the shape:

```
void setup() {
  size(120, 120);
```

```
}

void draw() {
  rotate(mouseX / 100.0);
  rect(-80, -10, 160, 20);
}
```

The previous pair of examples showed how to rotate around coordinate (0,0), but what about other possibilities? You can use the `translate()` and `rotate()` functions for more control. When they are combined, the order in which they appear affects the result. If the coordinate system is first moved and then rotated, that is different than first rotating the coordinate system, then moving it.

Example 6-5: Translation, then Rotation

To spin a shape around its center point at a place on screen away from the origin, first use `translate()` to move to the location where you'd like the shape, then call `rotate()`, and then draw the shape with its center at coordinate (0,0):

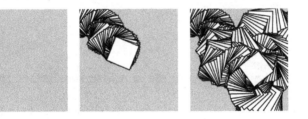

```
float angle = 0;

void setup() {
  size(120, 120);
}

void draw() {
  translate(mouseX, mouseY);
  rotate(angle);
  rect(-15, -15, 30, 30);
  angle += 0.1;
}
```

Example 6-6: Rotation, Then Translation

The following example is identical to Example 6-5 on page 79, except that translate() and rotate() are reversed. The shape now rotates around the upper-left corner of the Display Window, with the distance from the corner set by translate():

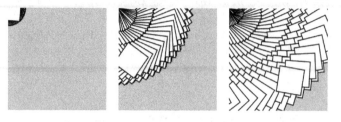

```
float angle = 0.0;

void setup() {
  size(120, 120);
}

void draw() {
  rotate(angle);
  translate(mouseX, mouseY);
  rect(-15, -15, 30, 30);
  angle += 0.1;
}
```

Another option is to use the rectMode(), ellipse Mode(), imageMode(), and shapeMode() functions, which make it easier to draw shapes from their center. You can read about these functions in the *Processing Reference*.

Example 6-7: An Articulating Arm

In this example, we've put together a series of translate() and rotate() functions to create a linked arm that bends back and forth. Each translate() further moves the position of the lines, and each rotate() adds to the previous rotation to bend more:

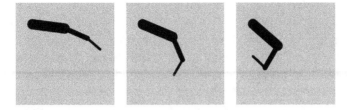

```
float angle = 0.0;
float angleDirection = 1;
float speed = 0.005;

void setup() {
  size(120, 120);
}

void draw() {
  background(204);
  translate(20, 25);   // Move to start position
  rotate(angle);
  strokeWeight(12);
  line(0, 0, 40, 0);
  translate(40, 0);    // Move to next joint
  rotate(angle * 2.0);
  strokeWeight(6);
  line(0, 0, 30, 0);
  translate(30, 0);    // Move to next joint
  rotate(angle * 2.5);
  strokeWeight(3);
  line(0, 0, 20, 0);

  angle += speed * angleDirection;
  if ((angle > QUARTER_PI) || (angle < 0)) {
    angleDirection = -angleDirection;
  }
}
```

The **angle** variable grows from 0 to **QUARTER_PI** (one quarter of the value of pi), then decreases until it is less than zero, then the cycle repeats. The value of the **angleDirection** variable is always 1 or –1 to make the value of **angle** correspondingly increase or decrease.

Scale

The scale() function stretches the coordinates on the screen. Because the coordinates expand or contract as the scale changes, everything drawn to the Display Window increases or decreases in dimension. Use scale(1.5) to make everything 150% of their original size, or scale(3) to make them three times larger. Using scale(1) would have no effect, because everything would remain 100% of the original. To make things half their size, use scale(0.5).

```
scale(1.5);                    scale(3);
rect(20, 20, 20, 40);          rect(20, 20, 20, 40);
```

Figure 6-3. *Scaling the coordinates*

Example 6-8: Scaling

Like rotate(), the scale() function transforms from the origin. Therefore, as with rotate(), to scale a shape from its center, translate to its location, scale, and then draw with the center at coordinate (0,0):

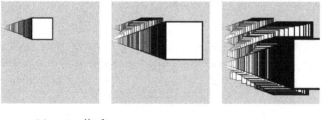

```
void setup() {
  size(120, 120);
}
```

```
void draw() {
  translate(mouseX, mouseY);
  scale(mouseX / 60.0);
  rect(-15, -15, 30, 30);
}
```

Example 6-9: Keeping Strokes Consistent

From the thick lines in Example 6-8 on page 82, you can see how the scale() function affects the stroke weight. To maintain a consistent stroke weight as a shape scales, divide the desired stroke weight by the scalar value:

```
void setup() {
  size(120, 120);
}

void draw() {
  translate(mouseX, mouseY);
  float scalar = mouseX / 60.0;
  scale(scalar);
  strokeWeight(1.0 / scalar);
  rect(-15, -15, 30, 30);
}
```

Push and Pop

To isolate the effects of a transformation so they don't affect later commands, use the pushMatrix() and popMatrix() functions. When pushMatrix() is run, it saves a copy of the current coordinate system and then restores that system after popMatrix(). This is useful when transformations are needed for one shape, but not wanted for another.

Example 6-10: Isolating Transformations

In this example, the smaller rectangle always draws in the same position because the `translate(mouseX, mouseY)` is cancelled by the `popMatrix()`:

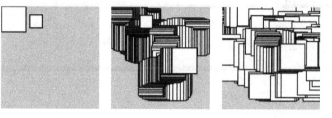

```
void setup() {
  size(120, 120);
}

void draw() {
  pushMatrix();
  translate(mouseX, mouseY);
  rect(0, 0, 30, 30);
  popMatrix();
  translate(35, 10);
  rect(0, 0, 15, 15);
}
```

The `pushMatrix()` and `popMatrix()` functions are always used in pairs. For every `pushMatrix()`, you need to have a matching `popMatrix()`.

Robot 4: Translate, Rotate, Scale

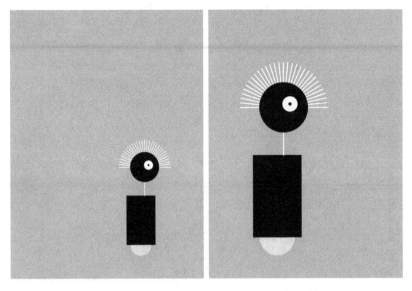

The `translate()`, `rotate()`, and `scale()` functions are all utilized in this modified robot sketch. In relation to "Robot 3: Response" on page 72, `translate()` is used to make the code easier to read. Here, notice how the x value no longer needs to be added to each drawing function because `translate()` moves everything.

Similarly, the `scale()` function is used to set the dimensions for the entire robot. When the mouse is not pressed, the size is set to 60%, and when it is pressed, it goes to 100% in relation to the original coordinates.

The `rotate()` function is used within a loop to draw a line, rotate it a little, then draw a second line, then rotate a little more, and so on until the loop has drawn 30 lines half-way around a circle to style a lovely head of robot hair:

```
float x = 60;           // x coordinate
float y = 440;          // y coordinate
int radius = 45;        // Head radius
int bodyHeight = 180;   // Body height
int neckHeight = 40;    // Neck height

float easing = 0.04;
```

```
void setup() {
  size(360, 480);
  ellipseMode(RADIUS);
}

void draw() {
  strokeWeight(2);

  float neckY = -1 * (bodyHeight + neckHeight + radius);

  background(0, 153, 204);

  translate(mouseX, y);  // Move all to (mouseX, y)

  if (mousePressed) {
    scale(1.0);
  } else {
    scale(0.6);  // 60% size when mouse is pressed
  }

  // Body
  noStroke();
  fill(255, 204, 0);
  ellipse(0, -33, 33, 33);
  fill(0);
  rect(-45, -bodyHeight, 90, bodyHeight-33);

  // Neck
  stroke(255);
  line(12, -bodyHeight, 12, neckY);

  // Hair
  pushMatrix();
  translate(12, neckY);
  float angle = -PI/30.0;
  for (int i = 0; i <= 30; i++) {
    line(80, 0, 0, 0);
    rotate(angle);
  }
  popMatrix();

  // Head
  noStroke();
  fill(0);
  ellipse(12, neckY, radius, radius);
  fill(255);
```

```
  ellipse(24, neckY-6, 14, 14);
  fill(0);
  ellipse(24, neckY-6, 3, 3);
}
```

7/Media

Processing is capable of drawing more than simple lines and shapes. It's time to learn how to load raster images, vector files, and fonts into our programs to extend the visual possibilities to photography, detailed diagrams, and diverse typefaces.

Processing uses a folder named *data* to store such files, so that you never have to think about their location when moving sketches around and exporting them. We've posted media files online for you to use in this chapter's examples: *http://www.processing.org/learning/books/media.zip*.

Download this file, unzip it to the desktop (or somewhere else convenient), and make a mental note of its location.

To unzip on Mac OS X, just double-click the file, and it will create a folder named *media*. On Windows, double-click the *media.zip* file, which will open a new window. In that window, drag the *media* folder to the desktop.

Create a new sketch, and select Add File from the Sketch menu. Find the *lunar.jpg* file from the *media* folder that you just unzipped and select it. If everything went well, the Message Area will read "One file added to the sketch."

To check for the file, select Show Sketch Folder in the Sketch menu. You should see a folder named *data*, with a copy of *lunar.jpg* inside. When you add a file to the sketch, the data folder will automatically be created. Instead of using the Add File menu command, you can do the same thing by dragging files into the editor area of the Processing window. The files will be copied to the *data* folder the same way (and the *data* folder will be created if none exists).

You can also create the *data* folder outside of Processing and copy files there yourself. You won't get the message saying that files have been added, but this is a helpful method when you're working with large numbers of files.

On Windows and Mac OS X, extensions are hidden by default. It's a good idea to change that option so that you always see the full name of your files. On Mac OS X, select Preferences from the Finder menu, and then make sure "Show all filename extensions" is checked in the Advanced tab. On Windows, look for Folder Options, and set the option there.

Images

There are three steps to follow before you can draw an image to the screen:

1. Add the image to the sketch's *data* folder (instructions given previously).
2. Create a PImage variable to store the image.
3. Load the image into the variable with loadImage().

Example 7-1: Load an Image

After all three steps are done, you can draw the image to the screen with the image() function. The first parameter to image() specifies the image to draw; the second and third set the x and y coordinates:

```
PImage img;

void setup() {
  size(480, 120);
  img = loadImage("lunar.jpg");
}

void draw() {
  image(img, 0, 0);
}
```

Optional fourth and fifth parameters set the width and height to draw the image. If the fourth and fifth parameters are not used, the image is drawn at the size at which it was created.

These next examples show how to work with more than one image in the same program and how to resize an image.

Example 7-2: Load More Images

For this example, you'll need to add the *capsule.jpg* file (found in the *media* folder you downloaded) to your sketch using one of the methods described earlier:

```
PImage img1;
PImage img2;

void setup() {
  size(480, 120);
  img1 = loadImage("lunar.jpg");
```

```
  img2 = loadImage("capsule.jpg");
}

void draw() {
  image(img1, -120, 0);
  image(img1, 130, 0, 240, 120);
  image(img2, 300, 0, 240, 120);
}
```

Example 7-3: Mousing Around with Images

When the mouseX and mouseY values are used as part of the fourth and fifth parameters of image(), the image size changes as the mouse moves:

```
PImage img;

void setup() {
  size(480, 120);
  img = loadImage("lunar.jpg");
}

void draw() {
  background(0);
  image(img, 0, 0, mouseX * 2, mouseY * 2);
}
```

When an image is displayed larger or smaller than its actual size, it may become distorted. Be careful to prepare your images at the sizes they will be used. When the display size of an image is changed with the image() function, the actual image on the hard drive doesn't change.

Processing can load and display raster images in the JPEG, PNG, and GIF formats. (Vector shapes in the SVG format can be displayed in a different way, as described in "Shapes" on page 97 later in this chapter.) You can convert images to the JPEG, PNG, and GIF formats using programs like GIMP and Photoshop. Most digital cameras save JPEG images that are much larger than the drawing area of most Processing sketches, so resizing such images before they are added to the *data* folder will make your sketches run more efficiently.

GIF and PNG images support transparency, which means that pixels can be invisible or partially visible (recall the discussion of color() and alpha values in Example 3-17 on page 26). GIF images have 1-bit transparency, which means that pixels are either fully opaque or fully transparent. PNG images have 8-bit transparency, which means that each pixel can have a variable level of opacity. The following examples show the difference, using the *clouds.gif* and *clouds.png* files found in the *media* folder that you downloaded. Be sure to add them to the sketch before trying each example.

Example 7-4: Transparency with a GIF

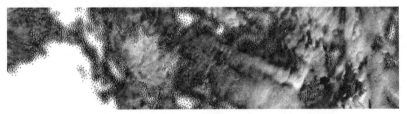

```
PImage img;

void setup() {
  size(480, 120);
  img = loadImage("clouds.gif");
}
void draw() {
  background(255);
  image(img, 0, 0);
  image(img, 0, mouseY * -1);
}
```

Example 7-5: Transparency with a PNG

```
PImage img;

void setup() {
  size(480, 120);
  img = loadImage("clouds.png");
}

void draw() {
  background(204);
  image(img, 0, 0);
  image(img, 0, mouseY * -1);
}
```

Remember to include the file extensions .*gif*, .*jpg*, or .*png* when you load the image. Also, be sure that the image name is typed exactly as it appears in the file, including the case of the letters. And if you missed it, read the note earlier in this chapter about making sure that the file extensions are visible on Mac OS X and Windows.

Fonts

The Processing software can display text using TrueType (.*ttf*) and OpenType (.*otf*) fonts, as well as a custom bitmap format called VLW. For this introduction, we will load a TrueType font from the *data* folder, the *SourceCodePro-Regular.ttf* font included in the *media* folder that you downloaded earlier.

The following websites are good places to find fonts with open licenses to use with Processing:

- Google Fonts (*http://www.google.com/fonts*)
- The Open Font Library (*http://openfontli brary.org*)
- The League of Moveable Type (*http:// www.theleagueofmoveabletype.com*)

Now it's possible to load the font and add words to a sketch. This part is similar to working with images, but there's one extra step:

1. Add the font to the sketch's *data* folder (instructions given previously).

2. Create a `PFont` variable to store the font.

3. Create the font and assign it to a variable with `createFont()`. This reads the font file, and creates a version of it at a specific size that can be used by Processing.

4. Use the `textFont()` function to set the current font.

Example 7-6: Drawing with Fonts

Now you can draw these letters to the screen with the `text()` function, and you can change the size with `textSize()`:

```
PFont font;

void setup() {
  size(480, 120);
  font = createFont("SourceCodePro-Regular.ttf", 32);
```

```
      textFont(font);
  }

  void draw() {
    background(102);
    textSize(32);
    text("That's one small step for man...", 25, 60);
    textSize(16);
    text("That's one small step for man...", 27, 90);
  }
```

The first parameter to text() is the character(s) to draw to the
screen. (Notice that the characters are enclosed within quotes.)
The second and third parameters set the horizontal and vertical
location. The location is relative to the baseline of the text (see
Figure 7-1).

(x,y)

Figure 7-1. *Typography coordinates*

Example 7-7: Draw Text in a Box

You can also set text to draw inside a box by adding fourth and
fifth parameters that specify the width and height of the box:

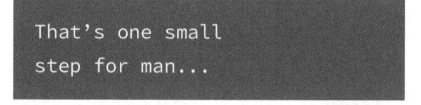

```
  PFont font;

  void setup() {
    size(480, 120);
```

```
  font = createFont("SourceCodePro-Regular.ttf", 24);
  textFont(font);
}

void draw() {
  background(102);
  text("That's one small step for man...", 26, 24, 240, 100);
}
```

Example 7-8: Store Text in a String

In the previous example, the words inside the text() function start to make the code difficult to read. We can store these words in a variable to make the code more modular. The String data type is used to store text data. Here's a new version of the previous example that uses a String:

```
PFont font;
String quote = "That's one small step for man...";

void setup() {
  size(480, 120);
  font = createFont("SourceCodePro-Regular.ttf", 24);
  textFont(font);
}

void draw() {
  background(102);
  text(quote, 26, 24, 240, 100);
}
```

There's a set of additional functions that affect how letters are displayed on screen. They are explained, with examples, in the Typography category of the *Processing Reference*.

Shapes

If you make vector shapes in a program like Inkscape or Illustrator, you can load them into Processing directly. This is helpful for shapes you'd rather not build with Processing's drawing functions. As with images, you need to add them to your sketch before they can be loaded.

There are three steps to load and draw an SVG file:

1. Add an SVG file to the sketch's *data* folder.
2. Create a PShape variable to store the vector file.
3. Load the vector file into the variable with loadShape().

Example 7-9: Draw with Shapes

After following these steps, you can draw the image to the screen with the shape() function:

```
PShape network;

void setup() {
  size(480, 120);
  network = loadShape("network.svg");
}

void draw() {
  background(0);
  shape(network, 30, 10);
  shape(network, 180, 10, 280, 280);
}
```

The parameters for shape() are similar to image(). The first parameter tells shape() which SVG to draw and the next pair sets the position. Optional fourth and fifth parameters set the width and height.

Example 7-10: Scaling Shapes

Unlike raster images, vector shapes can be scaled to any size without losing resolution. In this example, the shape is scaled based on the mouseX variable, and the shapeMode() function is used to draw the shape from its center, rather than the default position, the upper-left corner:

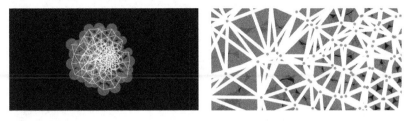

```
PShape network;

void setup() {
  size(240, 120);
  shapeMode(CENTER);
  network = loadShape("network.svg");
}

void draw() {
  background(0);
  float diameter = map(mouseX, 0, width, 10, 800);
  shape(network, 120, 60, diameter, diameter);
}
```

Processing doesn't support all SVG features. See the entry for **PShape** in the *Processing Reference* for more details.

Example 7-11: Creating a New Shape

In addition to loading shapes through the *data* folder, new shapes can be created with code through the `createShape()` function. In the next example, one of the creatures from Example 3-21 on page 29 is built in the `setup()` function. Once this happens, the shape can be used anywhere in the program with the `shape()` function:

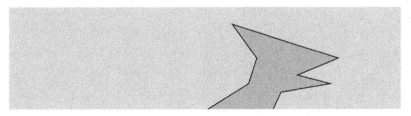

```
PShape dino;

void setup() {
  size(480, 120);
  dino = createShape();
  dino.beginShape();
  dino.fill(153, 176, 180);
  dino.vertex(50, 120);
  dino.vertex(100, 90);
  dino.vertex(110, 60);
  dino.vertex(80, 20);
  dino.vertex(210, 60);
  dino.vertex(160, 80);
  dino.vertex(200, 90);
  dino.vertex(140, 100);
  dino.vertex(130, 120);
  dino.endShape();
}

void draw() {
  background(204);
  translate(mouseX - 120, 0);
  shape(dino, 0, 0);
}
```

Making a custom **PShape** with **createShape()** can make sketches more efficient when the same shape is drawn many times.

Robot 5: Media

Unlike the robots created from lines and rectangles drawn in Processing in the previous chapters, these robots were created with a vector drawing program. For some shapes, it's often easier to point and click in a software tool like Inkscape or Illustrator than to define the shapes with coordinates in code.

There's a trade-off to selecting one image creation technique over another. When shapes are defined in Processing, there's more flexibility to modify them while the program is running. If the shapes are defined elsewhere and then loaded into Processing, changes are limited to the position, angle, and size. When loading each robot from an SVG file, as this example shows, the variations featured in Robot 2 (see "Robot 2: Variables" on page 47) are impossible.

Images can be loaded into a program to bring in visuals created in other programs or captured with a camera. With this image in the background, our robots are now exploring for life-forms in Norway at the dawn of the 20th century.

The SVG and PNG files used in this example can be downloaded from *http://www.processing.org/learning/books/media.zip*:

```
PShape bot1;
PShape bot2;
PShape bot3;
PImage landscape;

float easing = 0.05;
float offset = 0;

void setup() {
  size(720, 480);
  bot1 = loadShape("robot1.svg");
  bot2 = loadShape("robot2.svg");
  bot3 = loadShape("robot3.svg");
  landscape = loadImage("alpine.png");
}

void draw() {
  // Set the background to the "landscape" image, this image
  // must be the same width and height as the program
  background(landscape);

  // Set the left/right offset and apply easing to make
  // the transition smooth
  float targetOffset = map(mouseY, 0, height, -40, 40);
  offset += (targetOffset - offset) * easing;

  // Draw the left robot
  shape(bot1, 85 + offset, 65);

  // Draw the right robot smaller and give it a smaller offset
  float smallerOffset = offset * 0.7;
  shape(bot2, 510 + smallerOffset, 140, 78, 248);

  // Draw the smallest robot, give it a smaller offset
  smallerOffset *= -0.5;
  shape(bot3, 410 + smallerOffset, 225, 39, 124);
}
```

8/Motion

Like a flip book, animation on screen is created by drawing an image, then drawing a slightly different image, then another, and so on. The illusion of fluid motion is created by *persistence of vision*. When a set of similar images is presented at a fast enough rate, our brains translate these images into motion.

Frames

To create smooth motion, Processing tries to run the code inside **draw()** at 60 frames each second. A *frame* is one trip through the **draw()** and the *frame rate* is how many frames are drawn each second. Therefore, a program that draws 60 frames each second means the program runs the entire code inside **draw()** 60 times each second.

Example 8-1: See the Frame Rate

To confirm the frame rate, run this program and watch the values print to the Console (the **frameRate** variable keeps track of the program's speed):

```
void draw() {
  println(frameRate);
}
```

Example 8-2: Set the Frame Rate

The frameRate() function changes the speed at which the program runs. To see the result, uncomment different versions of frameRate() in this example:

```
void setup() {
  frameRate(30);      // Thirty frames each second
  //frameRate(12);    // Twelve frames each second
  //frameRate(2);     // Two frames each second
  //frameRate(0.5);   // One frame every two seconds
}

void draw() {
  println(frameRate);
}
```

Processing *tries* to run the code at 60 frames each second, but if it takes longer than 1/60th of a second to run the draw() method, then the frame rate will decrease. The frameRate() function specifies only the maximum frame rate, and the actual frame rate for any program depends on the computer that is running the code.

Speed and Direction

To create fluid motion examples, we use a data type called float. This type of variable stores numbers with decimal places, which provide more resolution for working with motion. For instance, when using int, the slowest you can move each frame is one pixel at a time (1, 2, 3, 4, . . .), but with float, you can move as slowly as you want (1.01, 1.01, 1.02, 1.03, . . .).

Example 8-3: Move a Shape

The following example moves a shape from left to right by updating the x variable:

```
int radius = 40;
float x = -radius;
float speed = 0.5;

void setup() {
  size(240, 120);
  ellipseMode(RADIUS);
}

void draw() {
  background(0);
  x += speed; // Increase the value of x
  arc(x, 60, radius, radius, 0.52, 5.76);
}
```

When you run this code, you'll notice the shape moves off the right of the screen when the value of the x variable is greater than the width of the window. The value of x continues to increase, but the shape is no longer visible.

Example 8-4: Wrap Around

There are many alternatives to this behavior, which you can choose from according to your preference. First, we'll extend the code to show how to move the shape back to the left edge of the screen after it disappears off the right. In this case, picture the screen as a flattened cylinder, with the shape moving around the outside to return to its starting point:

```
int radius = 40;
float x = -radius;
float speed = 0.5;

void setup() {
  size(240, 120);
  ellipseMode(RADIUS);
}

void draw() {
  background(0);
  x += speed;  // Increase the value of x
  if (x > width+radius) {  // If the shape is off screen,
    x = -radius;  // move to the left edge
  }
  arc(x, 60, radius, radius, 0.52, 5.76);
}
```

On each trip through **draw()**, the code tests to see if the value of x has increased beyond the width of the screen (plus the radius of the shape). If it has, we set the value of x to a negative value, so that as it continues to increase, it will enter the screen from the left. See Figure 8-1 for a diagram of how it works.

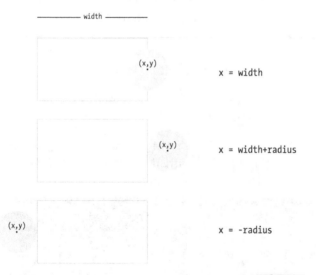

Figure 8-1. *Testing for the edges of the window*

Example 8-5: Bounce Off the Wall

In this example, we'll extend Example 8-3 on page 104 to have the shape change directions when it hits an edge, instead of wrapping around to the left. To make this happen, we add a new variable to store the direction of the shape. A direction value of 1 moves the shape to the right, and a value of –1 moves the shape to the left:

```
int radius = 40;
float x = 110;
float speed = 0.5;
int direction = 1;

void setup() {
  size(240, 120);
  ellipseMode(RADIUS);
}

void draw() {
  background(0);
  x += speed * direction;
  if ((x > width-radius) || (x < radius)) {
    direction = -direction; // Flip direction
  }
  if (direction == 1) {
    arc(x, 60, radius, radius, 0.52, 5.76); // Face right
  } else {
    arc(x, 60, radius, radius, 3.67, 8.9);  // Face left
  }
}
```

When the shape reaches an edge, this code flips the shape's direction by changing the sign of the **direction** variable. For example, if the **direction** variable is positive when the shape reaches an edge, the code flips it to negative.

Tweening

Sometimes you want to animate a shape to go from one point on screen to another. With a few lines of code, you can set up the start position and the stop position, then calculate the in-between (*tween*) positions at each frame.

Example 8-6: Calculate Tween Positions

To make this example code modular, we've created a group of variables at the top. Run the code a few times and change the values to see how this code can move a shape from any location to any other at a range of speeds. Change the step variable to alter the speed:

```
int startX = 20;        // Initial x coordinate
int stopX = 160;        // Final x coordinate
int startY = 30;        // Initial y coordinate
int stopY = 80;         // Final y coordinate
float x = startX;       // Current x coordinate
float y = startY;       // Current y coordinate
float step = 0.005;     // Size of each step (0.0 to 1.0)
float pct = 0.0;        // Percentage traveled (0.0 to 1.0)

void setup() {
  size(240, 120);
}

void draw() {
  background(0);
  if (pct < 1.0) {
    x = startX + ((stopX-startX) * pct);
    y = startY + ((stopY-startY) * pct);
    pct += step;
  }
```

```
    ellipse(x, y, 20, 20);
}
```

Random

Unlike the smooth, linear motion common to computer graphics, motion in the physical world is usually idiosyncratic. For instance, think of a leaf floating to the ground, or an ant crawling over rough terrain. We can simulate the unpredictable qualities of the world by generating random numbers. The `random()` function calculates these values; we can set a range to tune the amount of disarray in a program.

Example 8-7: Generate Random Values

The following short example prints random values to the Console, with the range limited by the position of the mouse. The `random()` function always returns a floating-point value, so be sure the variable on the left side of the assignment operator (=) is a `float` as it is here:

```
void draw() {
    float r = random(0, mouseX);
    println(r);
}
```

Example 8-8: Draw Randomly

The following example builds on Example 8-7 on page 109; it uses the values from `random()` to change the position of lines on screen. When the mouse is at the left of the screen, the change is small; as it moves to the right, the values from `random()` increase and the movement becomes more exaggerated. Because the `random()` function is inside the `for` loop, a new random value is calculated for each point of every line:

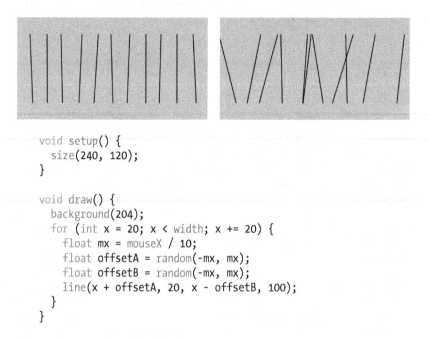

```
void setup() {
  size(240, 120);
}

void draw() {
  background(204);
  for (int x = 20; x < width; x += 20) {
    float mx = mouseX / 10;
    float offsetA = random(-mx, mx);
    float offsetB = random(-mx, mx);
    line(x + offsetA, 20, x - offsetB, 100);
  }
}
```

Example 8-9: Move Shapes Randomly

When used to move shapes around on screen, random values can generate images that are more natural in appearance. In the following example, the position of the circle is modified by random values on each trip through draw(). Because the back ground() function is not used, past locations are traced:

```
float speed = 2.5;
int diameter = 20;
float x;
float y;

void setup() {
  size(240, 120);
  x = width/2;
```

```
  y = height/2;
}

void draw() {
  x += random(-speed, speed);
  y += random(-speed, speed);
  ellipse(x, y, diameter, diameter);
}
```

If you watch this example long enough, you may see the circle leave the window and come back. This is left to chance, but we could add a few **if** structures or use the **constrain()** function to keep the circle from leaving the screen. The **constrain()** function limits a value to a specific range, which can be used to keep x and y within the boundaries of the Display Window. By replacing the **draw()** in the preceding code with the following, you'll ensure that the ellipse will remain on the screen:

```
void draw() {
  x += random(-speed, speed);
  y += random(-speed, speed);
  x = constrain(x, 0, width);
  y = constrain(y, 0, height);
  ellipse(x, y, diameter, diameter);
}
```

 The **randomSeed()** function can be used to force **random()** to produce the same sequence of numbers each time a program is run. This is described further in the *Processing Reference*.

Timers

Every Processing program counts the amount of time that has passed since it was started. It counts in milliseconds (thousandths of a second), so after 1 second, the counter is at 1,000; after 5 seconds, it's at 5,000; and after 1 minute, it's at 60,000. We can use this counter to trigger animations at specific times. The **millis()** function returns this counter value.

Example 8-10: Time Passes

You can watch the time pass when you run this program:

```
void draw() {
  int timer = millis();
  println(timer);
}
```

Example 8-11: Triggering Timed Events

When paired with an if block, the values from millis() can be used to sequence animation and events within a program. For instance, after two seconds have elapsed, the code inside the if block can trigger a change. In this example, variables called time1 and time2 determine when to change the value of the x variable:

```
int time1 = 2000;
int time2 = 4000;
float x = 0;

void setup() {
  size(480, 120);
}

void draw() {
  int currentTime = millis();
  background(204);
  if (currentTime > time2) {
    x -= 0.5;
  } else if (currentTime > time1) {
    x += 2;
  }
  ellipse(x, 60, 90, 90);
}
```

Circular

If you're a trigonometry ace, you already know how amazing the *sine* and *cosine* functions are. If you're not, we hope the next examples will trigger your interest. We won't discuss the math in detail here, but we'll show a few applications to generate fluid motion.

Figure 8-2 shows a visualization of sine wave values and how they relate to angles. At the top and bottom of the wave, notice how the rate of change (the change on the vertical axis) slows down, stops, then switches direction. It's this quality of the curve that generates interesting motion.

The sin() and cos() functions in Processing return values between –1 and 1 for the sine or cosine of the specified angle. Like arc(), the angles must be given in radian values (see Example 3-7 on page 18 and Example 3-8 on page 19 for a reminder of how radians work). To be useful for drawing, the float values returned by sin() and cos() are usually multiplied by a larger value.

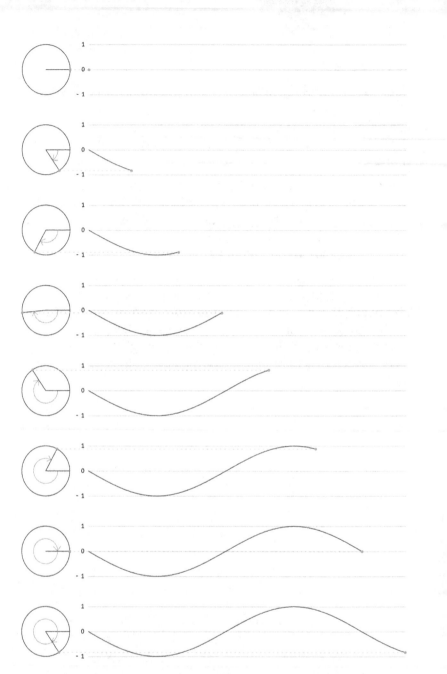

Figure 8-2. *A sine wave is created by tracing the sine values of an angle that moves around a circle*

Example 8-12: Sine Wave Values

This example shows how values for `sin()` cycle from −1 to 1 as the angle increases. With the `map()` function, the `sinval` variable is converted from this range to values from 0 and 255. This new value is used to set the background color of the window:

```
float angle = 0.0;

void draw() {
  float sinval = sin(angle);
  println(sinval);
  float gray = map(sinval, -1, 1, 0, 255);
  background(gray);
  angle += 0.1;
}
```

Example 8-13: Sine Wave Movement

This example shows how these values can be converted into movement:

```
float angle = 0.0;
float offset = 60;
float scalar = 40;
float speed = 0.05;

void setup() {
  size(240, 120);
}

void draw() {
  background(0);
  float y1 = offset + sin(angle) * scalar;
  float y2 = offset + sin(angle + 0.4) * scalar;
  float y3 = offset + sin(angle + 0.8) * scalar;
  ellipse( 80, y1, 40, 40);
  ellipse(120, y2, 40, 40);
```

```
    ellipse(160, y3, 40, 40);
    angle += speed;
  }
```

Example 8-14: Circular Motion

When sin() and cos() are used together, they can produce circular motion. The cos() values provide the x coordinates, and the sin() values provide the y coordinates. Both are multiplied by a variable named scalar to change the radius of the movement and summed with an offset value to set the center of the circular motion:

```
float angle = 0.0;
float offset = 60;
float scalar = 30;
float speed = 0.05;

void setup() {
  size(120, 120);
}

void draw() {
  float x = offset + cos(angle) * scalar;
  float y = offset + sin(angle) * scalar;
  ellipse( x, y, 40, 40);
  angle += speed;
}
```

Example 8-15: Spirals

A slight change made to increase the scalar value at each frame produces a spiral, rather than a circle:

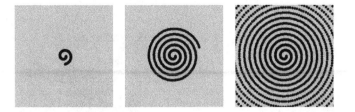

```
float angle = 0.0;
float offset = 60;
float scalar = 2;
float speed = 0.05;

void setup() {
  size(120, 120);
  fill(0);
}

void draw() {
  float x = offset + cos(angle) * scalar;
  float y = offset + sin(angle) * scalar;
  ellipse( x, y, 2, 2);
  angle += speed;
  scalar += speed;
}
```

Robot 6: Motion

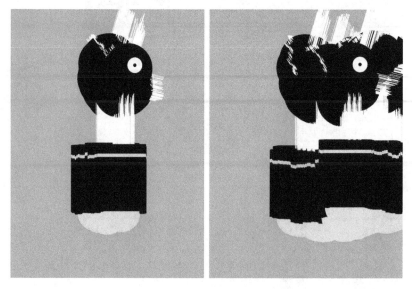

In this example, the techniques for random and circular motion are applied to the robot. The `background()` was removed to make it easier to see how the robot's position and body change.

At each frame, a random number between −4 and 4 is added to the *x* coordinate, and a random number between −1 and 1 is added to the *y* coordinate. This causes the robot to move more from left to right than top to bottom. Numbers calculated from the `sin()` function change the height of the neck so it oscillates between 50 and 110 pixels high:

```
float x = 180;            // x coordinate
float y = 400;            // y coordinate
float bodyHeight = 153;   // Body height
float neckHeight = 56;    // Neck height
float radius = 45;        // Head radius
float angle = 0.0;        // Angle for motion

void setup() {
  size(360, 480);
  ellipseMode(RADIUS);
  background(0, 153, 204);  // Blue background
}
```

```
void draw() {
  // Change position by a small random amount
  x += random(-4, 4);
  y += random(-1, 1);

  // Change height of neck
  neckHeight = 80 + sin(angle) * 30;
  angle += 0.05;

  // Adjust the height of the head
  float ny = y - bodyHeight - neckHeight - radius;

  // Neck
  stroke(255);
  line(x+2, y-bodyHeight, x+2, ny);
  line(x+12, y-bodyHeight, x+12, ny);
  line(x+22, y-bodyHeight, x+22, ny);

  // Antennae
  line(x+12, ny, x-18, ny-43);
  line(x+12, ny, x+42, ny-99);
  line(x+12, ny, x+78, ny+15);

  // Body
  noStroke();
  fill(255, 204, 0);
  ellipse(x, y-33, 33, 33);
  fill(0);
  rect(x-45, y-bodyHeight, 90, bodyHeight-33);
  fill(255, 204, 0);
  rect(x-45, y-bodyHeight+17, 90, 6);

  // Head
  fill(0);
  ellipse(x+12, ny, radius, radius);
  fill(255);
  ellipse(x+24, ny-6, 14, 14);
  fill(0);
  ellipse(x+24, ny-6, 3, 3);
}
```

9/Functions

Functions are the basic building blocks for Processing programs. They have appeared in every example we've presented. For instance, we've frequently used the **size()** function, the **line()** function, and the **fill()** function. This chapter shows how to write new functions to extend the capabilities of Processing beyond its built-in features.

The power of functions is modularity. Functions are independent software units that are used to build more complex programs—like LEGO bricks, where each type of brick serves a specific purpose, and making a complex model requires using the different parts together. As with functions, the true power of these bricks is the ability to build many different forms from the same set of elements. The same group of LEGOs that makes a spaceship can be reused to construct a truck, a skyscraper, and many other objects.

Functions are helpful if you want to draw a more complex shape like a tree over and over. The function to draw the tree shape would be made up of Processing's built-in fuctions, like **line()**, that create the form. After the code to draw the tree is written, you don't need to think about the details of tree drawing again—you can simply write **tree()** (or whatever you named the function) to draw the shape. Functions allow a complex sequence of statements to be abstracted, so you can focus on the higher-level goal (such as drawing a tree), and not the details of the implementation (the **line()** functions that define the tree shape). Once a function is defined, the code inside the function need not be repeated again.

Function Basics

A computer runs a program one line at a time. When a function is run, the computer jumps to where the function is defined and runs the code there, then jumps back to where it left off.

Example 9-1: Roll the Dice

This behavior is illustrated with the `rollDice()` function written for this example. When a program starts, it runs the code in `setup()` and then stops. The program takes a detour and runs the code inside `rollDice()` each time it appears:

```
void setup() {
  println("Ready to roll!");
  rollDice(20);
  rollDice(20);
  rollDice(6);
  println("Finished.");
}

void rollDice(int numSides) {
  int d = 1 + int(random(numSides));
  println("Rolling... " + d);
}
```

The two lines of code in `rollDice()` select a random number between 1 and the number of sides on the dice, and prints that number to the Console. Because the numbers are random, you'll see different numbers each time the program is run:

```
Ready to roll!
Rolling... 20
Rolling... 11
Rolling... 1
Finished.
```

Each time the `rollDice()` function is run inside `setup()`, the code within the function runs from top to bottom, then the program continues on the next line within `setup()`.

The `random()` function returns a number from 0 up to (but not including) the number specified. So `random(6)` returns a number between 0 and 5.99999. . . Because `random()` returns a float value, we also use `int()` to convert it to an integer. So `int(ran dom(6))` will return 0, 1, 2, 3, 4, or 5. Then we add 1 so that the number returned is between 1 and 6 (like a die). Like many other cases in this book, counting from 0 makes it easier to use the results of `random()` with other calculations.

Example 9-2: Another Way to Roll

If an equivalent program were written without the `rollDice()` function, it might look like this:

```
void setup() {
  println("Ready to roll!");
  int d1 = 1 + int(random(20));
  println("Rolling... " + d1);
  int d2 = 1 + int(random(20));
  println("Rolling... " + d2);
  int d3 = 1 + int(random(6));
  println("Rolling... " + d3);
  println("Finished.");
}
```

The `rollDice()` function in Example 9-1 on page 122 makes the code easier to read and maintain. The program is clearer, because the name of the function clearly states its purpose. In this example, we see the `random()` function in `setup()`, but its use is not as obvious. The number of sides on the die is also clearer with a function: when the code says `rollDice(6)`, it's obvious that it's simulating the roll of a six-sided die. Also, it's easier to maintain Example 9-1 on page 122, because information is not repeated. The phase `Rolling...` is repeated three times here. If you want to change that text to something else, you would need to update the program in three places, rather than making a single edit inside the `rollDice()` function. In addition, as you'll see in Example 9-5 on page 126, a function can also make a program much shorter (and therefore easier to maintain and read), which helps reduce the potential number of bugs.

Make a Function

In this section, we'll draw an owl to explain the steps involved in making a function.

Example 9-3: Draw the Owl

First, we'll draw the owl without using a function:

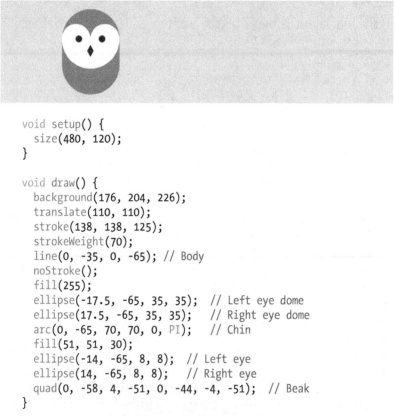

```
void setup() {
  size(480, 120);
}

void draw() {
  background(176, 204, 226);
  translate(110, 110);
  stroke(138, 138, 125);
  strokeWeight(70);
  line(0, -35, 0, -65); // Body
  noStroke();
  fill(255);
  ellipse(-17.5, -65, 35, 35);  // Left eye dome
  ellipse(17.5, -65, 35, 35);   // Right eye dome
  arc(0, -65, 70, 70, 0, PI);   // Chin
  fill(51, 51, 30);
  ellipse(-14, -65, 8, 8);  // Left eye
  ellipse(14, -65, 8, 8);   // Right eye
  quad(0, -58, 4, -51, 0, -44, -4, -51);  // Beak
}
```

Notice that **translate()** is used to move the origin (0,0) to 110 pixels over and 110 pixels down. Then the owl is drawn relative to (0,0), with its coordinates sometimes positive and negative as it's centered around the new 0,0 point. See Figure 9-1.

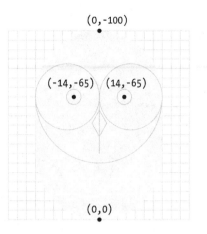

(0,-100)

(-14,-65)　(14,-65)

(0,0)

Figure 9-1. *The owl's coordinates*

Example 9-4: Two's Company

The code presented in Example 9-3 on page 124 is reasonable if there is only one owl, but when we draw a second, the length of the code is nearly doubled:

```
void setup() {
  size(480, 120);
}

void draw() {
  background(176, 204, 226);

  // Left owl
  translate(110, 110);
  stroke(138, 138, 125);
  strokeWeight(70);
  line(0, -35, 0, -65); // Body
  noStroke();
```

```
    fill(255);
    ellipse(-17.5, -65, 35, 35);   // Left eye dome
    ellipse(17.5, -65, 35, 35);    // Right eye dome
    arc(0, -65, 70, 70, 0, PI);    // Chin
    fill(51, 51, 30);
    ellipse(-14, -65, 8, 8);   // Left eye
    ellipse(14, -65, 8, 8);    // Right eye
    quad(0, -58, 4, -51, 0, -44, -4, -51); // Beak

    // Right owl
    translate(70, 0);
    stroke(138, 138, 125);
    strokeWeight(70);
    line(0, -35, 0, -65); // Body
    noStroke();
    fill(255);
    ellipse(-17.5, -65, 35, 35);   // Left eye dome
    ellipse(17.5, -65, 35, 35);    // Right eye dome
    arc(0, -65, 70, 70, 0, PI);    // Chin
    fill(51, 51, 30);
    ellipse(-14, -65, 8, 8);   // Left eye
    ellipse(14, -65, 8, 8);    // Right eye
    quad(0, -58, 4, -51, 0, -44, -4, -51); // Beak
}
```

The program grew from 21 lines to 34: the code to draw the first owl was cut and pasted into the program and a `translate()` was inserted to move it 70 pixels to the right. This is a tedious and inefficient way to draw a second owl, not to mention the headache of adding a third owl with this method. But duplicating the code is unnecessary, because this is the type of situation where a function can come to the rescue.

Example 9-5: An Owl Function

In this example, a function is introduced to draw two owls with the same code. If we make the code that draws the owl to the screen into a new function, the code need only appear once in the program:

```
void setup() {
  size(480, 120);
}

void draw() {
  background(176, 204, 226);
  owl(110, 110);
  owl(180, 110);
}

void owl(int x, int y) {
  pushMatrix();
  translate(x, y);
  stroke(138, 138, 125);
  strokeWeight(70);
  line(0, -35, 0, -65); // Body
  noStroke();
  fill(255);
  ellipse(-17.5, -65, 35, 35); // Left eye dome
  ellipse(17.5, -65, 35, 35);  // Right eye dome
  arc(0, -65, 70, 70, 0, PI);  // Chin
  fill(51, 51, 30);
  ellipse(-14, -65, 8, 8); // Left eye
  ellipse(14, -65, 8, 8);  // Right eye
  quad(0, -58, 4, -51, 0, -44, -4, -51); // Beak
  popMatrix();
}
```

You can see from the illustrations that this example and Example 9-4 on page 125 have the same result, but this example is shorter, because the code to draw the owl appears only once, inside the aptly named owl() function. This code runs twice, because it's called twice inside draw(). The owl is drawn in two different locations because of the parameters passed into the function that set the x and y coordinates.

Parameters are an important part of functions, because they provide flexibility. We saw another example in the rollDice()

function; the single parameter named `numSides` made it possible to simulate a 6-sided die, a 20-sided die, or a die with any number of sides. This is just like many other Processing functions. For instance, the parameters to the `line()` function make it possible to draw a line from any pixel on screen to any other pixel. Without the parameters, the function would be able to draw a line only from one fixed point to another.

Each parameter has a data type (such as `int` or `float`), because each parameter is a variable that's created each time the function runs. When this example is run, the first time the `owl` function is called, the value of the x parameter is 110, and y is also 110. In the second use of the function, the value of x is 180 and y is again 110. Each value is passed into the function and then wherever the variable name appears within the function, it's replaced with the incoming value.

Make sure the values passed into a function match the data types of the parameters. For instance, if the following appeared inside the `setup()` for this example:

```
owl(110.5, 120.2);
```

This would create an error, because the data type for the x and y parameters is `int`, and the values 110.5 and 120.2 are `float` values.

Example 9-6: Increasing the Surplus Population

Now that we have a basic function to draw the owl at any location, we can draw many owls efficiently by placing the function within a `for` loop and changing the first parameter each time through the loop:

```
void setup() {
  size(480, 120);
}

void draw() {
  background(176, 204, 226);
  for (int x = 35; x < width + 70; x += 70) {
    owl(x, 110);
  }
}
```

```
// Insert owl() function from Example 9-5
```

It's possible to keep adding more and more parameters to the function to change different aspects of how the owl is drawn. Values could be passed in to change the owl's color, rotation, scale, or the diameter of its eyes.

Example 9-7: Owls of Different Sizes

In this example, we've added two parameters to change the gray value and size of each owl:

```
void setup() {
  size(480, 120);
}

void draw() {
  background(176, 204, 226);
  randomSeed(0);
  for (int i = 35; i < width + 40; i += 40) {
    int gray = int(random(0, 102));
    float scalar = random(0.25, 1.0);
    owl(i, 110, gray, scalar);
  }
}

void owl(int x, int y, int g, float s) {
```

```
pushMatrix();
translate(x, y);
scale(s);  // Set the size
stroke(138-g, 138-g, 125-g); // Set the color value
strokeWeight(70);
line(0, -35, 0, -65); // Body
noStroke();
fill(255);
ellipse(-17.5, -65, 35, 35); // Left eye dome
ellipse(17.5, -65, 35, 35);  // Right eye dome
arc(0, -65, 70, 70, 0, PI);  // Chin
fill(51, 51, 30);
ellipse(-14, -65, 8, 8);  // Left eye
ellipse(14, -65, 8, 8);   // Right eye
quad(0, -58, 4, -51, 0, -44, -4, -51); // Beak
popMatrix();
}
```

Return Values

Functions can make a calculation and then return a value to the
main program. We've already used functions of this type, including random() and sin(). Notice that when this type of function
appears, the return value is usually assigned to a variable:

```
float r = random(1, 10);
```

In this case, random() returns a value between 1 and 10, which is
then assigned to the r variable.

A function that returns a value is also frequently used as a
parameter to another function. For instance:

```
point(random(width), random(height));
```

In this case, the values from random() aren't assigned to a variable—they are passed as parameters to point() and used to
position the point within the window.

Example 9-8: Return a Value

To make a function that returns a value, replace the keyword
void with the data type that you want the function to return. In
your function, specify the data to be passed back with the keyword return. For instance, this example includes a function

called `calculateMars()` that calculates the weight of a person or object on our neighboring planet:

```
void setup() {
  float yourWeight = 132;
  float marsWeight = calculateMars(yourWeight);
  println(marsWeight);
}

float calculateMars(float w) {
  float newWeight = w * 0.38;
  return newWeight;
}
```

Notice the data type `float` before the function name to show that it returns a floating-point value, and the last line of the block, which returns the variable `newWeight`. In the second line of `setup()`, that value is assigned to the variable `marsWeight`. (To see your own weight on Mars, change the value of the `your Weight` variable to your weight.)

Robot 7: Functions

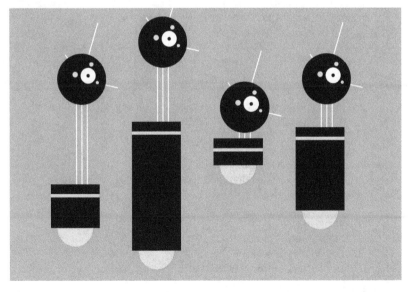

In contrast to Robot 2 (see "Robot 2: Variables" on page 47), this example uses a function to draw four robot variations within the same program. Because the `drawRobot()` function appears four times within `draw()`, the code within the `drawRobot()` block is run four times, each time with a different set of parameters to change the position and height of the robot's body.

Notice how what were global variables in Robot 2 have now been isolated within the `drawRobot()` function. Because these variables apply only to drawing the robot, they belong inside the curly braces that define the `drawRobot()` function block. Because the value of the `radius` variable doesn't change, it need not be a parameter. Instead, it is defined at the beginning of `drawRobot()`:

```
void setup() {
  size(720, 480);
  strokeWeight(2);
  ellipseMode(RADIUS);
}

void draw() {
  background(0, 153, 204);
  drawRobot(120, 420, 110, 140);
  drawRobot(270, 460, 260, 95);
  drawRobot(420, 310, 80, 10);
  drawRobot(570, 390, 180, 40);
}

void drawRobot(int x, int y, int bodyHeight, int neckHeight) {

  int radius = 45;
  int ny = y - bodyHeight - neckHeight - radius;   // Neck y

  // Neck
  stroke(255);
  line(x+2, y-bodyHeight, x+2, ny);
  line(x+12, y-bodyHeight, x+12, ny);
  line(x+22, y-bodyHeight, x+22, ny);

  // Antennae
  line(x+12, ny, x-18, ny-43);
  line(x+12, ny, x+42, ny-99);
  line(x+12, ny, x+78, ny+15);

  // Body
  noStroke();
  fill(255, 204, 0);
```

```
  ellipse(x, y-33, 33, 33);
  fill(0);
  rect(x-45, y-bodyHeight, 90, bodyHeight-33);
  fill(255, 204, 0);
  rect(x-45, y-bodyHeight+17, 90, 6);

  // Head
  fill(0);
  ellipse(x+12, ny, radius, radius);
  fill(255);
  ellipse(x+24, ny-6, 14, 14);
  fill(0);
  ellipse(x+24, ny-6, 3, 3);
  fill(153, 204, 255);
  ellipse(x, ny-8, 5, 5);
  ellipse(x+30, ny-26, 4, 4);
  ellipse(x+41, ny+6, 3, 3);
}
```

10/Objects

Object-oriented programming (OOP) is a different way to think about your programs. Although the term "object-oriented programming" may sound intimidating, there's good news: you've been working with objects since Chapter 7, when you started using **PImage**, **PFont**, **String**, and **PShape**. Unlike the primitive data types **boolean**, **int**, and **float**, which can store only one value, an object can store many. But that's only a part of the story. *Objects* are also a way to group variables with related functions. Because you already know how to work with variables and functions, objects simply combine what you've already learned into a more understandable package.

Objects are important, because they break up ideas into smaller building blocks. This mirrors the natural world where, for instance, organs are made of tissue, tissue is made of cells, and so on. Similarly, as your code becomes more complicated, you must think in terms of smaller structures that form more complicated ones. It's easier to write and maintain smaller, understandable pieces of code that work together than it is to write one large piece of code that does everything at once.

Fields and Methods

A software object is a collection of related variables and functions. In the context of objects, a variable is called a *field* (or *instance variable*) and a function is called a *method*. Fields and methods work just like the variables and functions covered in earlier chapters, but we'll use the new terms to emphasize that they are a part of an object. To say it another way, an object combines related data (fields) with related actions and behaviors (methods). The idea is to group together related data with related methods that act on that data.

For instance, to model a radio, think about what parameters can be adjusted and the actions that can affect those parameters:

Fields
    volume, frequency, band(FM, AM), power(on, off)

Methods
    setVolume, setFrequency, setBand

Modeling a simple mechanical device is easy compared to modeling an organism like an ant or a person. It's not possible to reduce such complex organisms to a few fields and methods, but it is possible to model enough to create an interesting simulation. *The Sims* video game is a clear example. This game is played by managing the daily activities of simulated people. The characters have enough personality to make a playable, addictive game, but no more. In fact, they have only five personality attributes: neat, outgoing, active, playful, and nice. With the knowledge that it's possible to make a highly simplified model of complex organisms, we could start programming an ant with only a few fields and methods:

Fields
    type(worker, soldier), weight, length

Methods
    walk, pinch, releasePheromones, eat

If you made a list of an ant's fields and methods, you might choose to focus on different aspects of the ant to model. There's no right way to make a model, as long as you make it appropriate for the purpose of your program's goals.

Define a Class

Before you can create an object, you must define a *class*. A class is the specification for an object. Using an architectural analogy, a class is like a blueprint for a house, and the object is like the house itself. Each house made from the blueprint can have variations, and the blueprint is only the specification, not a built structure. For example, one house can be blue and the other red; one house might come with a fireplace and the other without. Likewise with objects, the class defines the data types and behaviors, but each object (house) made from a single class (blueprint) has variables (color, fireplace) that are set to different values. To use a more technical term, each object is an *instance* of a class and each instance has its own set of fields and methods.

Before you write a class, we recommend a little planning. Think about what fields and methods your class should have. Do a little brainstorming to imagine all the possible options and then prioritize and make your best guess about what will work. You'll make changes during the programming process, but it's important to have a good start.

For your fields, select clear names and decide the data type for each. The fields inside a class can be any type of data. A class can simultaneously hold many images, boolean, float, and String values, and so on. Keep in mind that one reason to make a class is to group together related data elements. For your methods, select clear names and decide the return values (if any). The methods are used to change the values of the fields and to perform actions based on the fields' values.

For our first class, we'll convert Example 8-9 on page 110 from earlier in the book. We start by making a list of the fields from the example:

```
float x
float y
int diameter
float speed
```

The next step is to figure out what methods might be useful for the class. In looking at the draw() function from the example we're adapting, we see two primary components. The position of the shape is updated and drawn to the screen. Let's create two methods for our **class**, one for each task:

```
void move()
void display()
```

Neither of these methods return a value, so they both have the return type **void**. When we next write the **class** based on the lists of fields and methods, we'll follow four steps:

1. Create the block.
2. Add the fields.
3. Write a *constructor* (explained shortly) to assign values to the fields.
4. Add the methods.

First, we create a block:

```
class JitterBug {

}
```

Notice that the keyword **class** is lowercase and the name **Jitter Bug** is uppercase. Naming the **class** with an uppercase letter isn't required, but it is a convention (that we strongly encourage) used to denote that it's a **class**. (The keyword **class**, however, must be lowercase because it's a rule of the programming language.)

Second, we add the fields. When we do this, we have to decide which fields will have their values assigned through a *constructor*, a special method used for that purpose. As a rule of thumb, field values that you want to be different for each object are passed in through the constructor, and the other field values can be defined when they are declared. For the **JitterBug class**, we've decided that the values for **x**, **y**, and **diameter** will be passed in. So the fields are declared as follows:

```
class JitterBug {
  float x;
  float y;
  int diameter;
```

```
    float speed = 0.5;
  }
```

Third, we add the constructor. The constructor always has the same name as the **class**. The purpose of the constructor is to assign the initial values to the fields when an object (an instance of the **class**) is created (Figure 10-1). The code inside the constructor block is run once when the object is first created. As discussed earlier, we're passing in three parameters to the constructor when the object is initialized. Each of the values passed in is assigned to a temporary variable that exists only while the code inside the constructor is run. To clarify this, we've added the name *temp* to each of these variables, but they can be named with any terms that you prefer. They are used only to assign the values to the fields that are a part of the **class**. Also note that the constructor never returns a value and therefore doesn't have **void** or another data type before it. After adding the constructor, the **class** looks like this:

```
class JitterBug {

  float x;
  float y;
  int diameter;
  float speed = 0.5;

  JitterBug(float tempX, float tempY, int tempDiameter) {
    x = tempX;
    y = tempY;
    diameter = tempDiameter;
  }

}
```

The last step is to add the methods. This part is straightforward; it's just like writing functions, but here they are contained within the **class**. Also, note the code spacing. Every line within the **class** is indented a few spaces to show that it's inside the block. Within the constructor and the methods, the code is spaced again to clearly show the hierarchy:

```
class JitterBug {

  float x;
  float y;
```

```
int diameter;
float speed = 2.5;

JitterBug(float tempX, float tempY, int tempDiameter) {
  x = tempX;
  y = tempY;
  diameter = tempDiameter;
}

void move() {
  x += random(-speed, speed);
  y += random(-speed, speed);
}

void display() {
  ellipse(x, y, diameter, diameter);
}

}
```

```
Train red, blue;

void setup() {
  size(400, 400);
  red = new Train("Red Line", 90);
  blue = new Train("Blue Line", 120);
}

class Train {
  String name;
  int distance;
  Train (String tempName, int tempDistance) {
    name = tempName;
    distance = tempDistance;
  }
}
```

Assign "Red Line" to the name variable for the red object

Assign 90 to the distance variable for the red object

```
Train red, blue;

void setup() {
  size(400, 400);
  red = new Train("Red Line", 90);
  blue = new Train("Blue Line", 120);
}

class Train {
  String name;
  int distance;
  Train (String tempName, int tempDistance) {
    name = tempName;
    distance = tempDistance;
  }
}
```

Assign "Blue Line" to the name variable for the blue object

Assign 120 to the distance variable for the blue object

Figure 10-1. *Passing values into the constructor to set the values for an object's fields*

Create Objects

Now that you have defined a class, to use it in a program you must define an object from that class. There are two steps to create an object:

1. Declare the object variable.
2. Create (initialize) the object with the keyword new.

Example 10-1: Make an Object

To make your first object, we'll start by showing how this works within a Processing sketch and then continue by explaining each part in depth:

```
JitterBug bug;  // Declare object

void setup() {
  size(480, 120);
  // Create object and pass in parameters
  bug = new JitterBug(width/2, height/2, 20);
}

void draw() {
  bug.move();
  bug.display();
}

class JitterBug {

  float x;
  float y;
  int diameter;
  float speed = 2.5;

  JitterBug(float tempX, float tempY, int tempDiameter) {
```

```
    x = tempX;
    y = tempY;
    diameter = tempDiameter;
  }

  void move() {
    x += random(-speed, speed);
    y += random(-speed, speed);
  }

  void display() {
    ellipse(x, y, diameter, diameter);
  }

}
```

Each **class** is a *data type* and each object is a *variable*. We declare object variables in a similar way to variables from primitive data types like **boolean**, **int**, and **float**. The object is declared by stating the data type followed by a name for the variable:

```
JitterBug bug;
```

The second step is to initialize the object with the keyword **new**. It makes space for the object in memory and creates the fields. The name of the constructor is written to the right of the **new** keyword, followed by the parameters into the constructor, if any:

```
JitterBug bug = new JitterBug(200.0, 250.0, 30);
```

The three numbers within the parentheses are the parameters that are passed into the **JitterBug class** constructor. The number of these parameters and their data types must match those of the constructor.

Example 10-2: Make Multiple Objects

In Example 10-1 on page 142, we see something else new: the period (dot) that's used to access the object's methods inside of **draw()**. The dot operator is used to join the name of the object with its fields and methods. This becomes clearer in this example, where two objects are made from the same **class**. The **jit.move()** function refers to the **move()** method that belongs to

the object named **jit,** and **bug.move()** refers to the **move()**
method that belongs to the object named **bug:**

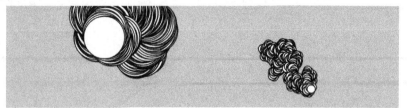

```
JitterBug jit;
JitterBug bug;

void setup() {
  size(480, 120);
  jit = new JitterBug(width * 0.33, height/2, 50);
  bug = new JitterBug(width * 0.66, height/2, 10);
}

void draw() {
  jit.move();
  jit.display();
  bug.move();
  bug.display();
}

class JitterBug {

  float x;
  float y;
  int diameter;
  float speed = 2.5;

  JitterBug(float tempX, float tempY, int tempDiameter) {
    x = tempX;
    y = tempY;
    diameter = tempDiameter;
  }

  void move() {
    x += random(-speed, speed);
    y += random(-speed, speed);
  }

  void display() {
```

```
    ellipse(x, y, diameter, diameter);
  }

}
```

Tabs

Now that the `class` exists as its own module of code, any changes will modify the objects made from it. For instance, you could add a field to the `JitterBug class` that controls the color, or another that determines its size. These values can be passed in using the constructor or assigned using additional methods, such as `setColor()` or `setSize()`. And because it's a self-contained unit, you can also use the `JitterBug class` in another sketch.

Now is a good time to learn about the tab feature of the Processing Development Environment (Figure 10-2). Tabs allow you to spread your code across more than one file. This makes longer code easier to edit and more manageable in general. A new tab is usually created for each `class`, which reinforces the modularity of working with classes and makes the code easy to find.

To create a new tab, click the arrow at the righthand side of the tab bar. When you select New Tab from the menu, you will be prompted to name the tab within the message window. Using this technique, modify this example's code to try to make a new tab for the `JitterBug class`.

Each tab shows up as a separate *.pde* file within the sketch's folder.

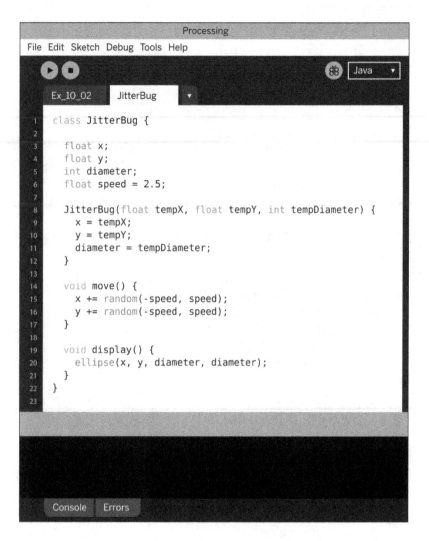

Figure 10-2. *Code can be split into different tabs to make it easier to manage*

Robot 8: Objects

A software object combines methods (functions) and fields (variables) into one unit. The **Robot class** in this example defines all of the robot objects that will be created from it. Each **Robot** object has its own set of fields to store a position and the illustration that will draw to the screen. Each has methods to update the position and display the illustration.

The parameters for **bot1** and **bot2** in **setup()** define the *x* and *y* coordinates and the *.svg* file that will be used to depict the robot. The **tempX** and **tempY** parameters are passed into the constructor and assigned to the **xpos** and **ypos** fields. The **svgName** parameter is used to load the related illustration. The objects (**bot1** and **bot2**) draw at their own location and with a different illustration because they each have unique values passed into the objects through their constructors:

```
Robot bot1;
Robot bot2;

void setup() {
  size(720, 480);
  bot1 = new Robot("robot1.svg", 90, 80);
  bot2 = new Robot("robot2.svg", 440, 30);
```

```
  }

void draw() {
  background(0, 153, 204);

  // Update and display first robot
  bot1.update();
  bot1.display();

  // Update and display second robot
  bot2.update();
  bot2.display();
}

class Robot {
  float xpos;
  float ypos;
  float angle;
  PShape botShape;
  float yoffset = 0.0;

  // Set initial values in constructor
  Robot(String svgName, float tempX, float tempY) {
    botShape = loadShape(svgName);
    xpos = tempX;
    ypos = tempY;
    angle = random(0, TWO_PI);
  }

  // Update the fields
  void update() {
    angle += 0.05;
    yoffset = sin(angle) * 20;
  }

  // Draw the robot to the screen
  void display() {
    shape(botShape, xpos, ypos + yoffset);
  }
}
```

11/Arrays

An *array* is a list of variables that share a common name. Arrays are useful because they make it possible to work with more variables without creating a new name for each. This makes the code shorter, easier to read, and more convenient to update.

From Variables to Arrays

When a program needs to keep track of one or two things, it's not necessary to use an array. In fact, adding an array might make the program more complicated than necessary. However, when a program has many elements (for example, a field of stars in a space game or multiple data points in a visualization), arrays make the code easier to write.

Example 11-1: Many Variables

To see what we mean, refer to Example 8-3 on page 104. This code works fine if we're moving around only one shape, but what if we want to have two? We need to make a new x variable and update it within **draw( )**:

```
float x1 = -20;
float x2 = 20;

void setup() {
  size(240, 120);
  noStroke();
}

void draw() {
  background(0);
  x1 += 0.5;
  x2 += 0.5;
  arc(x1, 30, 40, 40, 0.52, 5.76);
  arc(x2, 90, 40, 40, 0.52, 5.76);
}
```

Example 11-2: Too Many Variables

The code for the previous example is still manageable, but what if we want to have five circles? We need to add three more variables to the two we already have:

```
float x1 = -10;
float x2 = 10;
float x3 = 35;
float x4 = 18;
float x5 = 30;

void setup() {
  size(240, 120);
  noStroke();
}

void draw() {
  background(0);
  x1 += 0.5;
  x2 += 0.5;
  x3 += 0.5;
```

```
    x4 += 0.5;
    x5 += 0.5;
    arc(x1, 20, 20, 20, 0.52, 5.76);
    arc(x2, 40, 20, 20, 0.52, 5.76);
    arc(x3, 60, 20, 20, 0.52, 5.76);
    arc(x4, 80, 20, 20, 0.52, 5.76);
    arc(x5, 100, 20, 20, 0.52, 5.76);
}
```

This code is starting to get out of control.

Example 11-3: Arrays, Not Variables

Imagine what would happen if you wanted to have 3,000 circles. This would mean creating 3,000 individual variables, then updating each one separately. Could you keep track of that many variables? Would you want to? Instead, we use an array:

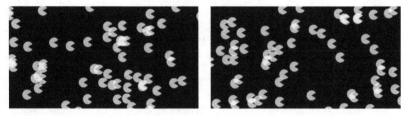

```
float[] x = new float[3000];

void setup() {
  size(240, 120);
  noStroke();
  fill(255, 200);
  for (int i = 0; i < x.length; i++) {
    x[i] = random(-1000, 200);
  }
}

void draw() {
  background(0);
  for (int i = 0; i < x.length; i++) {
    x[i] += 0.5;
    float y = i * 0.4;
    arc(x[i], y, 12, 12, 0.52, 5.76);
  }
}
```

We'll spend the rest of this chapter talking about the details that make this example possible.

Make an Array

Each item in an array is called an *element*, and each has an *index* value to mark its position within the array. Just like coordinates on the screen, index values for an array start counting from 0. For instance, the first element in the array has the index value 0, the second element in the array has the index value 1, and so on. If there are 20 values in the array, the index value of the last element is 19. Figure 11-1 shows the conceptual structure of an array.

```
int[] years = { 1920, 1972, 1980, 1996, 2010 };
```

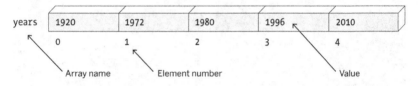

Figure 11-1. *An array is a list of one or more variables that share the same name*

Using arrays is similar to working with single variables; it follows the same patterns. As you know, you can make a single integer variable called x with this code:

```
int x;
```

To make an array, just place brackets after the data type:

```
int[] x;
```

The beauty of creating an array is the ability to make 2, 10, or 100,000 variable values with only one line of code. For instance, the following line creates an array of 2,000 integer variables:

```
int[] x = new int[2000];
```

You can make arrays from all Processing data types: `boolean`, `float`, `String`, `PShape`, and so on, as well as any user-defined `class`. For example, the following code creates an array of 32 `PImage` variables:

```
PImage[] images = new PImage[32];
```

To make an array, start with the name of the data type, followed by the brackets. The name you select for the array is next, followed by the assignment operator (the equal symbol), followed by the **new** keyword, followed by the name of the data type again, with the number of elements to create within the brackets. This pattern works for arrays of all data types.

Each array can store only one type of data (**boolean**, **int**, **float**, **PImage**, etc.). You can't mix and match different types of data within a single array. If you need to do this, work with objects instead.

Before we get ahead of ourselves, let's slow down and talk about working with arrays in more detail. Like making an object, there are three steps to working with an array:

1. Declare the array and define the data type.
2. Create the array with the keyword **new** and define the length.
3. Assign values to each element.

Each step can happen on its own line, or all the steps can be compressed together. Each of the three following examples shows a different technique to create an array called x that stores two integers, 12 and 2. Pay close attention to what happens before **setup()** and what happens within **setup()**.

Example 11-4: Declare and Assign an Array

First, we'll declare the array outside of **setup()** and then create and assign the values within. The syntax x[0] refers to the first element in the array and x[1] is the second:

```
int[] x;              // Declare the array

void setup() {
  size(200, 200);
  x = new int[2];    // Create the array
```

```
  x[0] = 12;        // Assign the first value
  x[1] = 2;         // Assign the second value
}
```

Example 11-5: Compact Array Assignment

Here's a slightly more compact example, in which the array is both declared and created on the same line, then the values are assigned within **setup()**:

```
int[] x = new int[2];   // Declare and create the array

void setup() {
  size(200, 200);
  x[0] = 12;        // Assign the first value
  x[1] = 2;         // Assign the second value
}
```

Example 11-6: Assigning to an Array in One Go

You can also assign values to the array when it's created, if it's all part of a single statement:

```
int[] x = { 12, 2 };   // Declare, create, and assign

void setup() {
  size(200, 200);
}
```

Avoid creating arrays within **draw()**, because creating a new array on every frame will slow down your frame rate.

Example 11-7: Revisiting the First Example

As a complete example of how to use arrays, we've recoded Example 11-1 on page 149 here. Although we don't yet see the full benefits revealed in Example 11-3 on page 151, we do see some important details of how arrays work:

```
float[] x = {-20, 20};

void setup() {
  size(240, 120);
  noStroke();
}

void draw() {
  background(0);
  x[0] += 0.5;  // Increase the first element
  x[1] += 0.5;  // Increase the second element
  arc(x[0], 30, 40, 40, 0.52, 5.76);
  arc(x[1], 90, 40, 40, 0.52, 5.76);
}
```

Repetition and Arrays

The for loop, introduced in "Repetition" on page 40, makes it easier to work with large arrays while keeping the code concise. The idea is to write a loop to move through each element of the array one by one. To do this, you need to know the length of the array. The length field associated with each array stores the number of elements. We use the name of the array with the dot operator (a period) to access this value. For instance:

```
int[] x = new int[2];    // Declare and create the array
println(x.length);       // Prints 2 to the Console

·int[] y = new int[1972];  // Declare and create the array
println(y.length);        // Prints 1972 to the Console
```

Example 11-8: Filling an Array in a for Loop

A for loop can be used to fill an array with values, or to read the values back out. In this example, the array is first filled with random numbers inside setup(), and then these numbers are used to set the stroke value inside draw(). Each time the program is run, a new set of random numbers is put into the array:

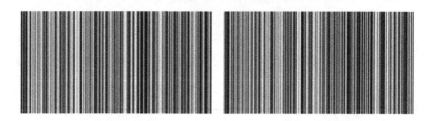

```
float[] gray;

void setup() {
  size(240, 120);
  gray = new float[width];
  for (int i = 0; i < gray.length; i++) {
    gray[i] = random(0, 255);
  }
}

void draw() {
  for (int i = 0; i < gray.length; i++) {
    stroke(gray[i]);
    line(i, 0, i, height);
  }
}
```

Example 11-9: Track Mouse Movements

In this example, there are two arrays to store the position of the mouse—one for the x coordinate and one for the y coordinate. These arrays store the location of the mouse for the previous 60 frames. With each new frame, the oldest x and y coordinate values are removed and replaced with the current mouseX and mouseY values. The new values are added to the first position of the array, but before this happens, each value in the array is moved one position to the right (from back to front) to make room for the new numbers. This example visualizes this action. Also, at each frame, all 60 coordinates are used to draw a series of ellipses to the screen:

```
int num = 60;
int[] x = new int[num];
int[] y = new int[num];

void setup() {
  size(240, 120);
  noStroke();
}

void draw() {
  background(0);
  // Copy array values from back to front
  for (int i = x.length-1; i > 0; i--) {
    x[i] = x[i-1];
    y[i] = y[i-1];
  }
  x[0] = mouseX;  // Set the first element
  y[0] = mouseY;  // Set the first element
  for (int i = 0; i < x.length; i++) {
    fill(i * 4);
    ellipse(x[i], y[i], 40, 40);
  }
}
```

The technique for storing a shifting buffer of numbers in an array shown in this example and Figure 11-2 is less efficient than an alternative technique that uses the % (modulo) operator. This is explained in the Examples → Basics → Input → StoringInput example included with Processing.

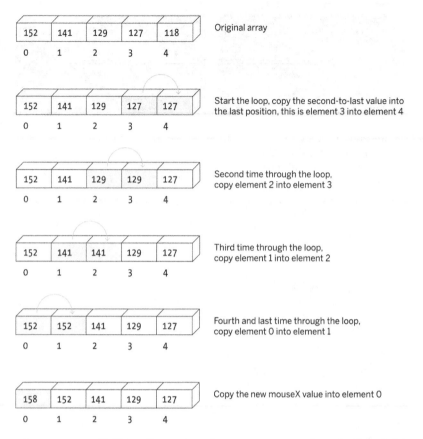

152	141	129	127	118
0	1	2	3	4

Original array

152	141	129	127	127
0	1	2	3	4

Start the loop, copy the second-to-last value into the last position, this is element 3 into element 4

152	141	129	129	127
0	1	2	3	4

Second time through the loop, copy element 2 into element 3

152	141	141	129	127
0	1	2	3	4

Third time through the loop, copy element 1 into element 2

152	152	141	129	127
0	1	2	3	4

Fourth and last time through the loop, copy element 0 into element 1

158	152	141	129	127
0	1	2	3	4

Copy the new mouseX value into element 0

Figure 11-2. *Shifting the values in an array one place to the right*

Arrays of Objects

The two short examples in this section bring together every major programming concept in this book: variables, iteration, conditionals, functions, objects, and arrays. Making an array of objects is nearly the same as making the arrays we introduced on the previous pages, but there's one additional consideration: because each array element is an object, it must first be created with the keyword **new** (like any other object) before it is assigned to the array. With a custom-defined **class** such as **JitterBug** (see Chapter 10), this means using **new** to set up each element before it's assigned to the array. Or, for a built-in Processing class such as PImage, it means using the loadImage() function to create the object before it's assigned.

Example 11-10: Managing Many Objects

This example creates an array of 33 JitterBug objects and then updates and displays each one inside draw(). For this example to work, you need to add the JitterBug class to the code:

```
JitterBug[] bugs = new JitterBug[33];

void setup() {
  size(240, 120);
  for (int i = 0; i < bugs.length; i++) {
    float x = random(width);
    float y = random(height);
    int r = i + 2;
    bugs[i] = new JitterBug(x, y, r);
  }
}

void draw() {
  for (int i = 0; i < bugs.length; i++) {
    bugs[i].move();
    bugs[i].display();
  }
}

// Insert JitterBug class from Example 10-1
```

Example 11-11: A New Way to Manage Objects

When working with arrays of objects, there's a different kind of loop to use called an "enhanced" for loop. Instead of creating a new counter variable, such as the i variable in Example 11-10 on page 159, it's possible to iterate over the elements of an array or list directly. In the following example, each object in the bugs array of JitterBug objects is assigned to b in order to run the move() and display() methods for all objects in the array.

The enhanced for loop is often tidier than looping with a number, although in this example, we didn't use it inside setup() because i was needed in two places inside the loop, demonstrating how sometimes it's helpful to have the number around:

```
JitterBug[] bugs = new JitterBug[33];

void setup() {
  size(240, 120);
  for (int i = 0; i < bugs.length; i++) {
    float x = random(width);
    float y = random(height);
    int r = i + 2;
    bugs[i] = new JitterBug(x, y, r);
  }
}

void draw() {
  for (JitterBug b : bugs) {
    b.move();
    b.display();
  }
}

// Insert JitterBug class from Example 10-1
```

The final array example loads a sequence of images and stores each as an element within an array of PImage objects.

Example 11-12: Sequences of Images

To run this example, get the images from the *media.zip* file as described in Chapter 7. The images are named sequentially (*frame-0000.png*, *frame-0001.png*, and so forth), which makes it possible to create the name of each file within a for loop, as seen in the eighth line of the program:

```
int numFrames = 12;  // The number of frames
PImage[] images = new PImage[numFrames];  // Make the array
int currentFrame = 0;

void setup() {
  size(240, 120);
  for (int i = 0; i < images.length; i++) {
    String imageName = "frame-" + nf(i, 4) + ".png";
    images[i] = loadImage(imageName);  // Load each image
  }
  frameRate(24);
}


void draw() {
  image(images[currentFrame], 0, 0);
  currentFrame++;        // Next frame
  if (currentFrame >= images.length) {
    currentFrame = 0;  // Return to first frame
  }
}
```

The nf() function formats numbers so that nf(1, 4) returns the string "0001" and nf(11, 4) returns "0011". These values are concatenated with the beginning of the filename (*frame-*) and the end (*.png*) to create the complete filename as a String variable. The files are loaded into the array on the following line. The images are displayed to the screen one at a time in draw(). When the last image in the array is displayed, the program returns to the beginning of the array and shows the images again in sequence.

Robot 9: Arrays

Arrays make it easier for a program to work with many elements. In this example, an array of Robot objects is declared at the top. The array is then allocated inside setup(), and each Robot object is created inside the for loop. In draw(), another for loop is used to update and display each element of the bots array.

The for loop and an array make a powerful combination. Notice the subtle differences between the code for this example and Robot 8 (see "Robot 8: Objects" on page 147) in contrast to the extreme changes in the visual result. Once an array is created and a for loop is put in place, it's as easy to work with 3 elements as it is 3,000.

The decision to load the SVG file within setup() rather than in the Robot class is the major change from Robot 8. This choice was made so the file is loaded only once, rather than as many times as there are elements in the array (in this case, 20 times). This change makes the code start faster because loading a file takes time, and it uses less memory because the file is stored once. Each element of the bot array references the same file:

```
Robot[] bots;   // Declare array of Robot objects

void setup() {
  size(720, 480);
  PShape robotShape = loadShape("robot2.svg");
  // Create the array of Robot objects
  bots = new Robot[20];
  // Create each object
  for (int i = 0; i < bots.length; i++) {
    // Create a random x coordinate
    float x = random(-40, width-40);
    // Assign the y coordinate based on the order
    float y = map(i, 0, bots.length, -100, height-200);
    bots[i] = new Robot(robotShape, x, y);
  }
}

void draw() {
  background(0, 153, 204);
  // Update and display each bot in the array
  for (int i = 0; i < bots.length; i++) {
    bots[i].update();
    bots[i].display();
  }
}

class Robot {
  float xpos;
  float ypos;
  float angle;
  PShape botShape;
  float yoffset = 0.0;

  // Set initial values in constructor
  Robot(PShape shape, float tempX, float tempY) {
    botShape = shape;
    xpos = tempX;
    ypos = tempY;
    angle = random(0, TWO_PI);
  }

  // Update the fields
  void update() {
    angle += 0.05;
    yoffset = sin(angle) * 20;
  }
```

```
// Draw the robot to the screen
void display() {
  shape(botShape, xpos, ypos + yoffset);
}
}
```

12/Data

Data visualization is one of the most active areas at the intersection of code and graphics and is also among the most popular uses of Processing. This chapter builds on what's already been discussed about storing and loading data earlier in the book and introduces more features relevant to data sets that can be used for visualization.

There's a wide range of software that can output standard visualizations like bar charts and scatter plots. However, writing code to create these visualizations from scratch provides more control over the output and encourages users to imagine, explore, and create more unique representations of data. For us, this is the point of learning to code using software like Processing, and we find it far more interesting than being limited by the prepackaged methods or tools that are available.

Data Summary

It's a good time to rewind and discuss how data was introduced throughout this book. A variable in a Processing sketch is used to store a piece of data. We started with primitives. In this case, the word *primitive* means a single piece of data. For instance, an int stores one whole number and cannot store more than one. The idea of *data types* is essential. Each kind of data is unique and is stored in a different way. Floating-point numbers (numbers with decimals), integers (no decimals), and alphanumeric symbols (letters and numbers) all have different *types* of data to store that kind of information like float, int, and char.

An array is created to store a list of elements within a single variable name. For instance, Example 11-8 on page 155 stores hundreds of floating-point numbers that are used to set the stroke value of lines. Arrays can be created from any primitive data type, but they are restricted to storing a single type of data. The way to store more than one data type within a single data structure is to create a `class`.

The `String`, `PImage`, `PFont`, and `PShape` classes store more than one data element and each is unique. For instance, a `String` can store more than one character, a word, sentence, paragraph, or more. In addition, it has methods to get the length of the data or return upper- or lowercase versions of it. As another example, a `PImage` has an array called `pixels` and variables that store the width and the height of the image.

Objects created from the `String`, `PImage`, and `PShape` classes can be defined within the code, but they can also be loaded from a file within a sketch's *data* folder. The examples in this chapter will also load data into sketches, but they utilize new classes that store data in different ways.

The `Table class` is introduced for storing data in a table of rows and columns. The `JSONObject` and `JSONArray` classes are introduced to store data loaded in through files that use the *JSON* format. The file formats for `Table`, `JSONObject`, and `JSONArray` are discussed in more detail in the following section.

The *XML* data format is another native data format for Processing and it's documented in the *Processing Reference*, but it's not covered in this text.

Tables

Many data sets are stored as rows and columns, so Processing includes a `Table class` to make it easier to work with tables. If you have worked with spreadsheets, you have a head start in working with tables in code. Processing can read a table from a file, or create a new one directly in code. It's also possible to read and write to any row or column and modify individual cells within the table. In this chapter, we will focus on working with table data.

Column   Cell coordinates (x,y)

0,0	1,0	2,0	3,0	
0,1	1,1	2,1	3,1	Row
0,2	1,2	2,2	3,2	
0,3	1,3	2,3	3,3	Cell

Figure 12-1. *Tables are grids of cells. Rows are the horizontal elements and columns are vertical. Data can be read from individual cells, rows, and columns.*

Table data is often stored in plain-text files with columns using commas or the tab character. A *comma-separated values* file is abbreviated as *CSV* and uses the file extension *.csv*. When tabs are used, the extension *.tsv* is sometimes used.

To load a CSV or TSV file, first place it into a sketch's *data* folder as described at the beginning of Chapter 7, and then use the loadTable() function to get the data and place it into an object made from the Table class.

Only the first few lines of each data set are shown in the following examples. If you're manually typing the code, you'll need the entire *.csv*, *.json*, or *.tsv* file to replicate the visualizations shown in the figures. You can get them from an example sketch's *data* folder (see "Examples and Reference" on page 11).

The data for the next example is a simplified version of Boston Red Sox player David Ortiz's batting statistics from 1997 to 2014. From left to right, it lists the year, number of home runs, runs batted in (RBIs), and batting average. When opened in a text editor, the first five lines of the file looks like this:

```
1997,1,6,0.327
1998,9,46,0.277
1999,0,0,0
```

```
2000,10,63,0.282
2001,18,48,0.234
```

Example 12-1: Read the Table

In order to load this data into Processing, an object from the Table class is created. The object in this example is called stats. The loadTable() function loads the *ortiz.csv* file from the *data* folder in the sketchbook. From there, the for loop reads through each table row in sequence. It gets the data from the table and saves it into int and float variables. The getRow Count() method is used to count the number of rows in the data file. Because this data is Ortiz's statistics from 1997 to 2014, there are 18 rows of data to read:

```
Table stats;

void setup() {
  stats = loadTable("ortiz.csv");
  for (int i = 0; i < stats.getRowCount(); i++) {
    // Gets an integer from row i, column 0 in the file
    int year = stats.getInt(i, 0);
    // Gets the integer from row i, column 1
    int homeRuns = stats.getInt(i, 1);
    int rbi = stats.getInt(i, 2);
    // Read a number that includes decimal points
    float average = stats.getFloat(i, 3);
    println(year, homeRuns, rbi, average);
  }
}
```

Inside the for loop, the getInt() and getFloat() methods are used to grab the data from the table. It's important to use the getInt() method to read integer data and likewise to use get Float() for floating-point variables. Both of these methods have two parameters, the first is the row to read from and the second is the column.

Example 12-2: Draw the Table

The next example builds on the last. It creates an array called homeRuns to store data after it is loaded inside setup() and the data from that array is used within draw(). The length of homeRuns is used three times with the code homeRuns.length, to count the number of for loop iterations.

homeRuns is used first in setup() to define how many times to get an integer from the table data. Second, it is used to place a vertical mark on the graph for each data item in the array. Third, it is used to read each element of the array one by one and to stop reading from the array at the end. After the data is loaded inside setup() and read into the array, the rest of this program applies what was learned in Chapter 11.

This example is the visualization of a simplified version of Boston Red Sox player David Ortiz's batting statistics from 1997 to 2014 drawn from a table:

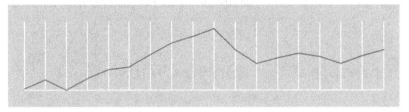

```
int[] homeRuns;

void setup() {
  size(480, 120);
  Table stats = loadTable("ortiz.csv");
  int rowCount = stats.getRowCount();
  homeRuns = new int[rowCount];
  for (int i = 0; i < homeRuns.length; i++) {
    homeRuns[i] = stats.getInt(i, 1);
  }
}

void draw() {
  background(204);
  // Draw background grid for data
  stroke(255);
  line(20, 100, 20, 20);
  line(20, 100, 460, 100);
  for (int i = 0; i < homeRuns.length; i++) {
    float x = map(i, 0, homeRuns.length-1, 20, 460);
    line(x, 20, x, 100);
  }
  // Draw lines based on home run data
```

```
noFill();
stroke(204, 51, 0);
beginShape();
for (int i = 0; i < homeRuns.length; i++) {
  float x = map(i, 0, homeRuns.length-1, 20, 460);
  float y = map(homeRuns[i], 0, 60, 100, 20);
  vertex(x, y);
}
endShape();
}
```

This example is so minimal that it's not necessary to store this data in arrays, but the idea can be applied to more complex examples you might want to make in the future. In addition, you can see how this example could be enhanced with more information—for instance, information on the vertical axis to state the number of home runs and on the horizontal to define the year.

Example 12-3: 29,740 Cities

To get a better idea about the potential of working with data tables, the next example uses a larger data set and introduces a convenient feature. This table data is different because the first row, the first line in the file, is a *header*. The header defines a label for each column to clarify the context. This is the first five lines of our new data file called *cities.csv*:

```
zip,state,city,lat,lng
35004,AL,Acmar,33.584132,-86.51557
35005,AL,Adamsville,33.588437,-86.959727
35006,AL,Adger,33.434277,-87.167455
35007,AL,Keystone,33.236868,-86.812861
```

The header makes it easier to read the code—for example, the second line of the file states the zip code of Acmar, Alabama, is 35004 and defines the latitude of the city as 33.584132 and the longitude as -86.51557. In total, the file is 29,741 lines long and it defines the location and zip codes of 29,740 cities in the United States.

The next example loads this data within **setup()** and then draws it to the screen in a **for** loop within **draw()**. The **setXY()** function converts the latitude and longitude data from the file into a point on the screen:

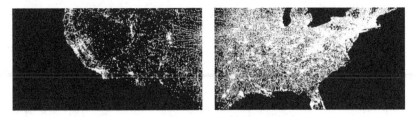

```
Table cities;

void setup() {
  size(240, 120);
  cities = loadTable("cities.csv", "header");
  stroke(255);
}

void draw() {
  background(0, 26, 51);
  float xoffset = map(mouseX, 0, width, -width*3, -width);
  translate(xoffset, -300);
  scale(10);
  strokeWeight(0.1);
  for (int i = 0; i < cities.getRowCount(); i++) {
    float latitude = cities.getFloat(i, "lat");
    float longitude = cities.getFloat(i, "lng");
    setXY(latitude, longitude);
  }
}

void setXY(float lat, float lng) {
  float x = map(lng, -180, 180, 0, width);
  float y = map(lat, 90, -90, 0, height);
  point(x, y);
}
```

Within setup(), notice a second parameter "header" is added to loadTable(). If this is not done, the code will treat the first line of the CSV file as data and not as the title of each column.

The Table class has dozens of methods for features like adding and removing columns and rows, getting a list of unique entries in a column, or sorting the table. A more complete list of methods along with short examples is included in the *Processing Reference*.

JSON

The JSON (JavaScript Object Notation) format is another common system for storing data. Like HTML and XML formats, the elements have labels associated with them. For instance, the data for a film could include labels for the title, director, release year, rating, and more.

These labels will be paired with the data like this:

```
"title": "Alphaville"
"director": "Jean-Luc Godard"
"year": 1964
"rating": 7.2
```

To work as a JSON file, the film labels need a little more punctuation to separate the elements. Commas are used between each data pair and braces enclose it. The data defined within the curly braces is a *JSON object*.

With these changes, our valid JSON data file looks like this:

```
{
  "title": "Alphaville",
  "director": "Jean-Luc Godard",
  "year": 1964,
  "rating": 7.2
}
```

There's another interesting detail in this short JSON sample related to data types: you'll notice that the title and director data is contained within quotes to mark them as **String** data and the year and rating are without quotes to define them as numbers. Specifically, the year is an integer and the rating is a floating-point number. This distinction becomes important after the data is loaded into a sketch.

To add another film to the list, a set of brackets placed at the top and bottom are used to signify that the data is an array of JSON objects. Each object is separated by a comma.

Putting it together looks like this:

```
[
  {
    "title": "Alphaville",
    "director": "Jean-Luc Godard",
    "year": 1964,
    "rating": 7.2
  },
  {
    "title": "Pierrot le Fou",
    "director": "Jean-Luc Godard",
    "year": 1965,
    "rating": 7.7
  }
]
```

This pattern can be repeated to include more films. At this point, it's interesting to compare this JSON notation to the corresponding table representation of the same data.

As a CSV file, the data looks like this:

```
title, director, year, rating
Alphaville, Jean-Luc Godard, 1964, 9.1
Pierrot le Fou, Jean-Luc Godard, 1965, 7.7
```

Notice that the CSV notation has fewer characters, which can be important when working with massive data sets. On the other hand, the JSON version is often easier to read because each piece of data is labeled.

Now that the basics of JSON and its relation to a Table has been introduced, let's look at the code needed to read a JSON file into a Processing sketch.

Example 12-4: Read a JSON File

This sketch loads the JSON file seen at the beginning of this section, the file that includes only the data for the single film *Alphaville*:

```
JSONObject film;

void setup() {
  film = loadJSONObject("film.json");
  String title = film.getString("title");
  String dir = film.getString("director");
  int year = film.getInt("year");
```

```
  float rating = film.getFloat("rating");
  println(title + " by " + dir + ", " + year);
  println("Rating: " + rating);
}
```

The JSONObject class is used to create a code object to store the data. Once that data is loaded, each individual piece of data can be read in sequence or by requesting the data related to each label. Notice that different data types are requested by the name of the data type. The getString() method is used for the name of the film and the getInt() method is used for the release year.

Example 12-5: Visualize Data from a JSON File

To work with the JSON file that includes more than one film, we need to introduce a new class, the JSONArray. Here, the data file started in the previous example has been updated to include all of the director's films from 1960–1966. The name of each film is placed in order on screen according to the release year and assigned a gray value based on the rating value.

There are several differences between this example and Example 12-4 on page 173. The most important is the way the JSON file is loaded into Film objects. The JSON file is loaded within setup() and each JSONObject that represents a single film is passed into the constructor of the Film class. The constructor assigns the JSONObject data to String, float, and int fields inside each Film object. The Film class also has a method to display the name of the film:

```
Film[] films;

void setup() {
  size(480, 120);
```

```
JSONArray filmArray = loadJSONArray("films.json");
films = new Film[filmArray.size()];
for (int i = 0; i < films.length; i++) {
  JSONObject o = filmArray.getJSONObject(i);
  films[i] = new Film(o);
}
}

void draw() {
  background(0);
  for (int i = 0; i < films.length; i++) {
    int x = i*32 + 32;
    films[i].display(x, 105);
  }
}

class Film {
  String title;
  String director;
  int year;
  float rating;

  Film(JSONObject f) {
    title = f.getString("title");
    director = f.getString("director");
    year = f.getInt("year");
    rating = f.getFloat("rating");
  }

  void display(int x, int y) {
    float ratingGray = map(rating, 6.5, 8.1, 102, 255);
    pushMatrix();
    translate(x, y);
    rotate(-QUARTER_PI);
    fill(ratingGray);
    text(title, 0, 0);
    popMatrix();
  }
}
```

This example is bare bones in its visualization of the film data. It shows how to load the data and how to draw based on those data values, but it's your challenge to format it to accentuate what you find interesting about the data. For example, is it more interesting to show the number of films Godard made each year? Is it more interesting to compare and contrast this data with the films of another director? Will all of this be easier to read with a different font, sketch size, or aspect ratio? The skills introduced in the earlier chapters in this book can be applied to bring this sketch to the next step of refinement.

Network Data and APIs

Public access to massive quantities of data collected by governments, corporations, organizations, and individuals is changing our culture, from the way we socialize to how we think about intangible ideas like privacy. This data is most often accessed through software structures called *APIs*.

The acronym *API* is mysterious and its meaning—application programming interface—isn't much clearer. However, APIs are essential for working with data and they aren't necessarily difficult to understand. Essentially, they are requests for data made to a service. When data sets are huge, it's not practical or desired to copy the entirety of the data; an API allows a programmer to request only the trickle of data that is relevant from a massive sea.

This concept can be more clearly illustrated with a hypothetical example. Let's assume there's an organization that maintains a database of temperature ranges for every city within a country. The API for this data set allows a programmer to request the high and low temperatures for any city during the month of October in 1972. In order to access this data, the request must be made through a specific line or lines of code, in the format mandated by the data service.

Some APIs are entirely public, but many require authentication, which is typically a unique user ID or key so the data service can keep track of its users. Most APIs have rules about how many, or how frequently requests can be made. For instance, it might be possible to make only 1,000 requests per month, or no more than one request per second.

Processing can request data over the Internet when the computer running the program is online. CSV, TSV, JSON, and XML files can be loaded using the corresponding load function with a

URL as the parameter. For instance, the current weather in Cincinnati is available in JSON format (*http://bit.ly/cin-json*).

Read the URL closely to decode it:

1. It requests data from the *api* subdomain of the *openweathermap.org* site.
2. It specifies a city to search for (*q* is an abbreviation for *query*, and is frequently used in URLs that specify searches).
3. It also indicates that the data will be returned in imperial format, meaning the temperature will be in Fahrenheit. Replacing *imperial* with *metric* will provide temperature data in degrees Celsius.

Looking at this data from OpenWeatherMap is a more realistic example of working with data found in the wild rather than the simplified data sets introduced earlier. At the time of this writing, the file returned from that URL looks like this:

```
{"message":"accurate","cod":"200","count":1,"list":[{"id":
4508722,"name":"Cincinnati","coord":{"lon":-84.456886,"lat":
39.161999},"main":{"temp":34.16,"temp_min":34.16,"temp_max":
34.16,"pressure":999.98,"sea_level":1028.34,"grnd_level":
999.98,"humidity":77},"dt":1423501526,"wind":{"speed":
9.48,"deg":354.002},"sys":{"country":"US"},"clouds":{"all":
80},"weather":[{"id":803,"main":"Clouds","description":"broken
clouds","icon":"04d"}]}]}
```

This file is much easier to read when it's formatted with line breaks, and the *JSON object* and array structures defined with braces and brackets:

```
{
  "message": "accurate",
  "count": 1,
  "cod": "200",
  "list": [{
    "clouds": {"all": 80},
    "dt": 1423501526,
    "coord": {
      "lon": -84.456886,
      "lat": 39.161999
    },
    "id": 4508722,
    "wind": {
      "speed": 9.48,
```

```
        "deg": 354.002
    },
    "sys": {"country": "US"},
    "name": "Cincinnati",
    "weather": [{
        "id": 803,
        "icon": "04d",
        "description": "broken clouds",
        "main": "Clouds"
    }],
    "main": {
        "humidity": 77,
        "pressure": 999.98,
        "temp_max": 34.16,
        "sea_level": 1028.34,
        "temp_min": 34.16,
        "temp": 34.16,
        "grnd_level": 999.98
    }
  }]
}
```

Note that brackets are seen in the **"list"** and **"weather"** sections, indicating an array of *JSON objects*. Although the array in this example only contains a single item, in other cases, the API might return multiple days or variations of the data from multiple weather stations.

Example 12-6: Parsing the Weather Data

The first step in working with this data is to study it and then to write minimal code to extract the desired data. In this case, we're curious about the current temperature. We can see that our temperature data is 34.16. It's labeled as temp and it's inside the main object, which is inside the list array. A function called getTemp() was written for this example to work with the format of this specific JSON file organization:

```
void setup() {
  float temp = getTemp("cincinnati.json");
  println(temp);
}

float getTemp(String fileName) {
```

```
JSONObject weather = loadJSONObject(fileName);
JSONArray list = weather.getJSONArray("list");
JSONObject item = list.getJSONObject(0);
JSONObject main = item.getJSONObject("main");
float temperature = main.getFloat("temp");
return temperature;
}
```

The name of the JSON file, *cincinnati.json*, is passed into the get
Temp() function inside setup() and loaded there. Next, because
of the format of the JSON file, a series of JSONArray and JSONOb
ject files are needed to get deeper and deeper into the data
structure to finally arrive at our desired number. This number is
stored in the temperature variable and then returned by the
function to be assigned to the temp variable in setup() where it is
printed to the Console.

Example 12-7: Chaining Methods

The sequence of JSON variables created in succession in the
last example can be approached differently by chaining the get
methods together. This example works like Example 12-6 on
page 178, but the methods are connected with the dot operator,
rather than calculated one at a time and assigned to objects in
between:

```
void setup() {
  float temp = getTemp("cincinnati.json");
  println(temp);
}

float getTemp(String fileName) {
  JSONObject weather = loadJSONObject(fileName);
  return weather.getJSONArray("list").getJSONObject(0).
  getJSONObject("main").getFloat("temp");
}
```

Also note how the final temperature value is returned by the
getTemp() function. In Example 12-6 on page 178, a float variable is created to store the decimal value, then that value is
returned. Here, the data created by the get methods is returned
directly, without intermediate variables.

This example can be modified to access more of the data from
the feed and to build a sketch that displays the data to the

screen rather than just writing it to the Console. You can also modify it to read data from another online API—you'll find that the data returned by many APIs shares a similar format.

Robot 10: Data

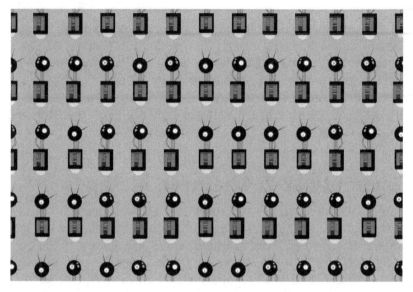

The final robot example in this book is different from the rest because it has two parts. The first part generates a data file using random values and **for** loops and the second part reads that data file to draw an army of robots onto the screen.

The first sketch uses two new code elements, the **PrintWriter** class and the **createWriter()** function. Used together, they create and open a file in the *sketchbook* folder to store the data generated by the sketch. In this example, the object created from **PrintWriter** is called **output** and the file is called *botArmy.tsv*. In the loops, data is written into the file by running the **println()** method on the output object. Here, random values are used to define which of three robot images will be drawn for each coordinate. For the file to be correctly created, the **flush()** and **close()** methods must be run before the program is stopped.

The code that draws an ellipse is a visual preview to reveal the location of the coordinate on screen, but notice that the ellipse isn't recorded into the file:

```
PrintWriter output;

void setup() {
  size(720, 480);
  // Create the new file
  output = createWriter("botArmy.tsv");
  // Write a header line with the column titles
  output.println("type\ttx\tty");
  for (int y = 0; y <= height; y += 120) {
    for (int x = 0; x <= width; x += 60) {
      int robotType = int(random(1, 4));
      output.println(robotType + "\t" + x + "\t" + y);
      ellipse(x, y, 12, 12);
    }
  }
  output.flush(); // Write the remaining data to the file
  output.close(); // Finish the file
}
```

After that program is run, open the *botArmy.tsv* file in the *sketchbook* folder to see how the data is written. The first five lines of that file will be similar to this:

type	x	y
3	0	0
1	20	0
2	40	0
1	60	0
3	80	0

The first column is used to define which robot image to use, the second column is the *x* coordinate, and the third column is the *y* coordinate.

The next sketch loads the *botArmy.tsv* file and uses the data for these purposes:

```
Table robots;
PShape bot1;
PShape bot2;
PShape bot3;

void setup() {
  size(720, 480);
```

```
background(0, 153, 204);
bot1 = loadShape("robot1.svg");
bot2 = loadShape("robot2.svg");
bot3 = loadShape("robot3.svg");
shapeMode(CENTER);
robots = loadTable("botArmy.tsv", "header");
for (int i = 0; i < robots.getRowCount(); i++) {
  int bot = robots.getInt(i, "type");
  int x = robots.getInt(i, "x");
  int y = robots.getInt(i, "y");
  float sc = 0.3;
  if (bot == 1) {
    shape(bot1, x, y, bot1.width*sc, bot1.height*sc);
  } else if (bot == 2) {
    shape(bot2, x, y, bot2.width*sc, bot2.height*sc);
  } else {
    shape(bot3, x, y, bot3.width*sc, bot3.height*sc);
  }
}
}
```

A more concise (and flexible) variation of this sketch uses arrays and the rows() method of the Table class as a more advanced approach:

```
int numRobotTypes = 3;
PShape[] shapes = new PShape[numRobotTypes];
float scalar = 0.3;

void setup() {
  size(720, 480);
  background(0, 153, 204);
  for (int i = 0; i < numRobotTypes; i++) {
    shapes[i] = loadShape("robot" + (i+1) + ".svg");
  }
  shapeMode(CENTER);
  Table botArmy = loadTable("botArmy.tsv", "header");
  for (TableRow row : botArmy.rows()) {
    int robotType = row.getInt("type");
    int x = row.getInt("x");
    int y = row.getInt("y");
    PShape bot = shapes[robotType - 1];
    shape(bot, x, y, bot.width*scalar, bot.height*scalar);
  }
}
```

13/Extend

This book focuses on using Processing for interactive graphics, because that's the core of what Processing does. However, the software can do much more and is often part of projects that move beyond a single computer screen. For example, Processing has been used to control machines, create images used in feature films, and export models for 3D printing.

Over the last decade, Processing has been used to make music videos for Radiohead and R.E.M., to create illustrations for publications such as *Nature* and the *New York Times*, to output sculptures for gallery exhibitions, to control huge video walls, to knit sweaters, and much more. Processing has this flexibility because of its system of libraries.

A Processing *library* is a collection of code that extends the software beyond its core functions and classes. Libraries have been important to the growth of the project, because they let developers add new features quickly. As smaller, self-contained projects, libraries are easier to manage than if these features were integrated into the main software.

To use a library, select Import Library from the Sketch menu and select the library you want to use from the list. Choosing a library adds a line of code that indicates that the library will be used with the current sketch.

For instance, when the PDF Export Library (pdf) is added, this line of code is added to the top of the sketch:

```
import processing.pdf.*;
```

In addition to the libraries included with Processing (these are called the *core* libraries), there are over 100 *contributed* libraries that are linked from the Processing website. All libraries are listed online at *http://processing.org/reference/libraries/*.

Before a contributed library can be imported through the Sketch menu, it must be added through the *Library Manager*. Select the Import Library option from the Sketchbook menu and then select Add Library to open the Library Manager interface. Click a library description and then click the Install button to download it to your computer.

The downloaded files are saved to the *libraries* folder that is located in your sketchbook. You can find the location of your sketchbook by opening the Preferences. The Library Manager can also be used to update and remove libraries.

As mentioned before, there are more than 100 Processing libraries, so they clearly can't all be discussed here. We've selected a few that we think are fun and useful to introduce in this chapter.

Sound

The *Sound audio library* introduced with Processing 3.0 has the ability to play, analyze, and generate (synthesize) sound. This library needs to be downloaded with the Library Manager as described earlier. (It's not included with the main Processing download because of its size.)

Like the images, shape files, and fonts introduced in Chapter 7, a *sound file* is another type of media to augment a Processing sketch. Processing's Sound library can load a range of file formats including WAV, AIFF, and MP3. Once a sound file is loaded, it can be played, stopped, and looped, or even distorted using different "effects" classes.

Example 13-1: Play a Sample

The most common use of the Sound library is to play a sound as background music or when an event happens on screen. The following example builds on Example 8-5 on page 107 to play a sound when the shape hits the edges of the screen. The *blip.wav* file is included in the *media* folder that you downloaded in Chapter 7 from *http://www.processing.org/learning/books/ media.zip*.

As with other media, the `SoundFile` object is defined at the top of the sketch, it's loaded within `setup()`, and after that, it can be used anywhere in the program:

```
import processing.sound.*;

SoundFile blip;

int radius = 120;
float x = 0;
float speed = 1.0;
int direction = 1;

void setup() {
  size(440, 440);
  ellipseMode(RADIUS);
  blip = new SoundFile(this, "blip.wav");
  x = width/2; // Start in the center
}

void draw() {
  background(0);
  x += speed * direction;
  if ((x > width-radius) || (x < radius)) {
    direction = -direction; // Flip direction
    blip.play();
  }
  if (direction == 1) {
    arc(x, 220, radius, radius, 0.52, 5.76); // Face right
  } else {
    arc(x, 220, radius, radius, 3.67, 8.9); // Face left
  }
}
```

The sound is triggered each time its `play()` method is run. This example works well because the sound is only played when the value of the x variable is at the edges of the screen. If the sound were played each time through `draw()`, the sound would restart 60 times each second and wouldn't have time to finish playing. The result is a rapid clipping sound. To play a longer sample while a program runs, call the `play()` or `loop()` method for that sound inside `setup()` so the sound is triggered only a single time.

The `SoundFile class` has many methods to control how a sound is played. The most essential are `play()` to play the sample a single time, `loop()` to play it from beginning to end over and over, `stop()` to halt the playback, and `jump()` to move to a specific moment within the file.

Example 13-2: Listen to a Microphone

In addition to playing a sound, Processing can *listen*. If your computer has a microphone, the Sound library can read live audio through it. Sounds from the mic can be analyzed, modified, and played:

```
import processing.sound.*;

AudioIn mic;
Amplitude amp;

void setup() {
```

```
    size(440, 440);
    background(0);
    // Create an audio input and start it
    mic = new AudioIn(this, 0);
    mic.start();
    // Create a new amplitude analyzer and patch into input
    amp = new Amplitude(this);
    amp.input(mic);
}


void draw() {
    // Draw a background that fades to black
    noStroke();
    fill(26, 76, 102, 10);
    rect(0, 0, width, height);
    // The analyze() method returns values between 0 and 1,
    // so map() is used to convert the values to larger numbers
    float diameter = map(amp.analyze(), 0, 1, 10, width);
    // Draw the circle based on the volume
    fill(255);
    ellipse(width/2, height/2, diameter, diameter);
}
```

There are two parts to getting the *amplitude* (volume) from an attached microphone. The AudioIn class is used to get the signal data from the mic and the Amplitude class is used to measure the signal. Objects from both classes are defined at the top of the code and created inside setup().

After the Amplitude object (named amp here) is made, the AudioIn object (named mic) is patched into the amp object with the input() method. After that, the analyze() method of the amp object can be run at any time to read the amplitude of the microphone data within the program. In this example, that is done each time through draw() and that value is then used to set the size of the circle.

In addition to playing a sound and analyzing sound as demonstrated in the last two examples, Processing can synthesize sound directly. The fundamentals of sound synthesis are waveforms that include the sine wave, triangle wave, and square wave.

A sine wave sounds smooth, a square wave is harsh, and a triangle wave is somewhere between. Each wave has a number of properties. The *frequency*, measured in hertz, determines the pitch, the highness or lowness of the tone. The *amplitude* of the wave determines the volume, the degree of loudness.

Example 13-3: Create a Sine Wave

In the following example, the value of `mouseX` determines the frequency of a sine wave. As the mouse moves left and right, the audible frequency and corresponding wave visualization increase and decrease:

```
import processing.sound.*;

SinOsc sine;

float freq = 400;

void setup() {
  size(440, 440);
  // Create and start the sine oscillator
  sine = new SinOsc(this);
  sine.play();
}

void draw() {
  background(176, 204, 176);
  // Map the mouseX value from 20Hz to 440Hz for frequency
  float hertz = map(mouseX, 0, width, 20.0, 440.0);
  sine.freq(hertz);
```

```
// Draw a wave to visualize the frequency of the sound
stroke(26, 76, 102);
for (int x = 0; x < width; x++) {
  float angle = map(x, 0, width, 0, TWO_PI * hertz);
  float sinValue = sin(angle) * 120;
  line(x, 0, x, height/2 + sinValue);
}
}
```

The **sine** object, created from the **SinOsc class**, is defined at the top of the code and then created inside **setup()**. Like working with a sample, the wave needs to be played with the **play()** method to start generating the sound. Within **draw()**, the **freq()** method continuously sets the frequency of the waveform based on the left-right position of the mouse.

Image and PDF Export

The animated images created by a Processing program can be turned into a file sequence with the **saveFrame()** function. When **saveFrame()** appears at the end of **draw()**, it saves a numbered sequence of TIFF-format images of the program's output named *screen-0001.tif*, *screen-0002.tif*, and so on, to the sketch's folder.

These files can be imported into a video or animation program and saved as a movie file. You can also specify your own file-name and image file format with a line of code like this:

```
saveFrame("output-####.png");
```

Use the **#** (hash mark) symbol to show where the numbers will appear in the filename. They are replaced with the actual frame numbers when the files are saved. You can also specify a sub-folder to save the images into, which is helpful when working with many image frames:

```
saveFrame("frames/output-####.png");
```

When using **saveFrame()** inside **draw()**, a new file is saved each frame—so watch out, as this can quickly fill your *sketch* folder with thousands of files.

Example 13-4: Saving Images

This example shows how to save images by storing enough frames for a two-second animation. It loads and moves the robot file from "Robot 5: Media" on page 101. See Chapter 7 for instructions for downloading the file *robot1.svg* and adding it to the sketch.

The example runs the program at 30 frames per second and then exits after 60 frames:

```
PShape bot;
float x = 0;

void setup() {
  size(720, 480);
  bot = loadShape("robot1.svg");
  frameRate(30);
}

void draw() {
  background(0, 153, 204);
  translate(x, 0);
  shape(bot, 0, 80);
  saveFrame("frames/SaveExample-####.tif");
  x += 12;

  if (frameCount > 60) {
    exit();
  }
}
```

Processing will write an image based on the file extension that you use (*.png*, *.jpg*, or *.tif* are all built in, and some platforms may support others). To retrieve the saved files, go to Sketch → Show Sketch Folder.

A *.tif* image is saved uncompressed, which is fast but takes up a lot of disk space. Both *.png* and *.jpg* will create smaller files, but because of the compression they will usually require more time to save, making the sketch run slowly.

If your desired output is vector graphics, you can write the output to PDF files for higher resolution. The PDF Export library makes it possible to write PDF files directly from a sketch. These vector graphics files can be scaled to any size without losing resolution, which makes them ideal for print output—from posters and banners to entire books.

Example 13-5: Draw to a PDF

This example builds on Example 13-4 on page 190 to draw more robots, but it removes the motion. The PDF library is imported at the top of the sketch to extend Processing to be able to write PDF files.

This sketch creates a PDF file called *Ex-13-5.pdf* because of the third and fourth parameters to `size()`:

```
import processing.pdf.*;

PShape bot;

void setup() {
  size(600, 800, PDF, "Ex-13-5.pdf");
  bot = loadShape("robot1.svg");
}

void draw() {
  background(0, 153, 204);
  for (int i = 0; i < 100; i++) {
    float rx = random(-bot.width, width);
    float ry = random(-bot.height, height);
    shape(bot, rx, ry);
  }
  exit();
}
```

The geometry is not drawn on the screen; it is written directly into the PDF file, which is saved into the sketch's folder. The code in this example runs once and then exits at the end of `draw()`. The resulting output is shown in Figure 13-1.

There are more PDF Export examples included with the Processing software. Look in the *PDF Export* (pdf) section of the Processing examples to see more techniques.

Figure 13-1. *PDF export from Example 3-5*

Hello, Arduino

Arduino is an electronics prototyping platform with a series of microcontroller boards and the software to program them. Processing and Arduino share a long history together; they are sister projects with many similar ideas and goals, though they address separate domains. Because they share the same editor and programming environment and a similar syntax, it's easy to move between them and to transfer knowledge about one into the other.

In this section, we focus on reading data into Processing from an Arduino board and then visualize that data on screen. This makes it possible to use new inputs into Processing programs and to allow Arduino programmers to see their sensor input as graphics. These new inputs can be anything that attaches to an Arduino board. These devices range from a distance sensor to a compass or a mesh network of temperature sensors.

This section assumes that you have an Arduino board and that you already have a basic working knowledge of how to use it. If not, you can learn more online at *http://www.arduino.cc* and in the excellent book *Getting Started with Arduino* by Massimo Banzi (Maker Media). Once you've covered the basics, you can learn more about sending data between Processing and Arduino in another outstanding book, *Making Things Talk* by Tom Igoe (Maker Media).

Data can be transferred between a Processing sketch and an Arduino board with some help from the Processing Serial Library. *Serial* is a data format that sends one *byte* at a time. In the world of Arduino, a `byte` is a data type that can store values between 0 and 255; it works like an `int`, but with a much smaller range. Larger numbers are sent by breaking them into a list of bytes and then reassembling them later.

In the following examples, we focus on the Processing side of the relationship and keep the Arduino code simple. We visualize the data coming in from the Arduino board one `byte` at a time. With the techniques covered in this book and the hundreds of Arduino examples online, we hope this will be enough to get you started.

Example 13-6: Read a Sensor

The following Arduino code is used with the next three Processing examples:

```
// Note: This is code for an Arduino board, not Processing

int sensorPin = 0;  // Select input pin
int val = 0;

void setup() {
  Serial.begin(9600);  // Open serial port
}

void loop() {
  val = analogRead(sensorPin) / 4;  // Read value from sensor
  Serial.write((byte)val);  // Print variable to serial port
  delay(100);  // Wait 100 milliseconds
}
```

There are two important details to note about this Arduino example. First, it requires attaching a sensor into the analog input on pin 0 on the Arduino board. You might use a light sensor (also called a photo resistor, photocell, or light-dependent resistor) or another analog resistor such as a thermistor (temperature-sensitive resistor), flex sensor, or pressure sensor (force-sensitive resistor). The circuit diagram and drawing of the breadboard with components are shown in Figure 13-2. Next, notice that the value returned by the analogRead() function is divided by 4 before it's assigned to val. The values from analog Read() are between 0 and 1023, so we divide by 4 to convert them to the range of 0 to 255 so that the data can be sent in a single byte.

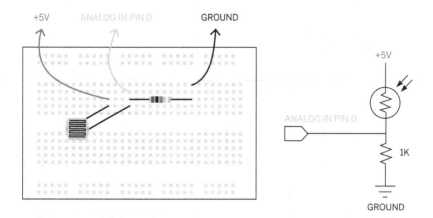

Figure 13-2. *Attaching a light sensor (photo resistor) to analog in pin 0*

Example 13-7: Read Data from the Serial Port

The first visualization example shows how to read the serial data in from the Arduino board and how to convert that data into the values that fit to the screen dimensions:

```
import processing.serial.*;

Serial port;  // Create object from Serial class
float val;  // Data received from the serial port

void setup() {
  size(440, 220);
  // IMPORTANT NOTE:
  // The first serial port retrieved by Serial.list()
  // should be your Arduino. If not, uncomment the next
  // line by deleting the // before it. Run the sketch
  // again to see a list of serial ports. Then, change
  // the 0 in between [ and ] to the number of the port
  // that your Arduino is connected to.
  //printArray(Serial.list());
  String arduinoPort = Serial.list()[0];
  port = new Serial(this, arduinoPort, 9600);
}
```

```
void draw() {
  if (port.available() > 0) {      // If data is available,
    val = port.read();            // read it and store it in val
    val = map(val, 0, 255, 0, height); // Convert the value
  }
  rect(40, val-10, 360, 20);
}
```

The Serial library is imported on the first line and the serial port is opened in **setup()**. It may or may not be easy to get your Processing sketch to talk with the Arduino board; it depends on your hardware setup. There is often more than one device that the Processing sketch might try to communicate with. If the code doesn't work the first time, read the comment in **setup()** carefully and follow the instructions.

Within **draw()**, the value is brought into the program with the **read()** method of the Serial object. The program reads the data from the serial port only when a new **byte** is available. The **available()** method checks to see if a new **byte** is ready and returns the number of bytes available. This program is written so that a single new **byte** will be read each time through **draw()**. The **map()** function converts the incoming value from its initial range from 0 to 255 to a range from 0 to the height of the screen; in this program, it's from 0 to 220.

Example 13-8: Visualizing the Data Stream

Now that the data is coming through, we'll visualize it in a more interesting format. The values coming in directly from a sensor are often erratic, and it's useful to smooth them out by averaging them. Here, we present the raw signal from the light sensor illustrated in the top half of the example and the smoothed signal in the bottom half:

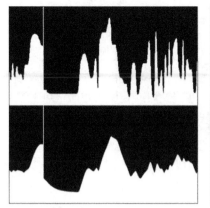
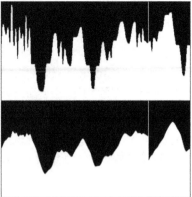

```
import processing.serial.*;

Serial port;   // Create object from Serial class
float val;     // Data received from the serial port
int x;
float easing = 0.05;
float easedVal;

void setup() {
  size(440, 440);
  frameRate(30);
  String arduinoPort = Serial.list()[0];
  port = new Serial(this, arduinoPort, 9600);
  background(0);
}

void draw() {
  if ( port.available() > 0) { // If data is available,
    val = port.read();          // read it and store it in val
    val = map(val, 0, 255, 0, height/2); // Convert the values
  }
  float targetVal = val;
  easedVal += (targetVal - easedVal) * easing;

  stroke(0);
  line(x, 0, x, height);           // Black line
  stroke(255);
  line(x+1, 0, x+1, height);       // White line
  line(x, 220, x, val);            // Raw value
  line(x, 440, x, easedVal + 220); // Averaged value

  x++;
```

```
  if (x > width) {
    x = 0;
  }
}
```

Similar to Example 5-8 on page 54 and Example 5-9 on page 55, this sketch uses the easing technique. Each new `byte` from the Arduino board is set as the target value, the difference between the current value and the target value is calculated, and the current value is moved closer to the target. Adjust the `easing` variable to affect the amount of smoothing applied to the incoming values.

Example 13-9: Another Way to Look at the Data

This example is inspired by radar display screens. The values are read in the same way from the Arduino board, but they are visualized in a circular pattern using the `sin()` and `cos()` functions introduced earlier in Example 8-12 on page 115, Example 8-13 on page 115, and Example 8-15 on page 116:

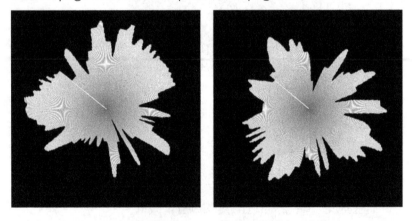

```
import processing.serial.*;

Serial port;    // Create object from Serial class
float val;      // Data received from the serial port
float angle;
float radius;

void setup() {
```

```
    size(440, 440);
    frameRate(30);
    strokeWeight(2);
    String arduinoPort = Serial.list()[0];
    port = new Serial(this, arduinoPort, 9600);
    background(0);
}

void draw() {
    if ( port.available() > 0) {  // If data is available,
      val = port.read();          // read it and store it in val
      // Convert the values to set the radius
      radius = map(val, 0, 255, 0, height * 0.45);
    }

    int middleX = width/2;
    int middleY = height/2;
    float x = middleX + cos(angle) * height/2;
    float y = middleY + sin(angle) * height/2;
    stroke(0);
    line(middleX, middleY, x, y);

    x = middleX + cos(angle) * radius;
    y = middleY + sin(angle) * radius;
    stroke(255);
    line(middleX, middleY, x, y);

    angle += 0.01;
}
```

The angle variable is updated continuously to move the line drawing the current value around the circle, and the val variable scales the length of the moving line to set its distance from the center of the screen. After one time around the circle, the values begin to write on top of the previous data.

We're excited about the potential of using Processing and Arduino together to bridge the world of software and electronics. Unlike the examples printed here, the communication can be bidirectional. Elements on screen can also affect what's happening on the Arduino board. This means you can use a Processing program as an interface between your computer and motors, speakers, lights, cameras, sensors, and almost anything else that can be controlled with an electrical signal. Again, check out *http://www.arduino.cc* for more information about Arduino.

A/Coding Tips

Coding is a type of writing. Like all types of writing, code has specific rules. For comparison, we'll quickly mention some of the rules for English that you probably haven't thought about in a while because they are second nature. Some of the more invisible rules are writing from left to right and putting a space between each word. More overt rules are spelling conventions, capitalizing the names of people and places, and using punctuation at the end of sentences to provide emphasis! If we break one or more of these rules when writing an email to a friend, the message still gets through. For example, "hello ben. how r u today" communicates nearly as well as, "Hello, Ben. How are you today?" However, flexibility with the rules of writing don't transfer to programming. Because you're writing to communicate with a computer, rather than another person, you need to be more precise and careful. One misplaced character is often the difference between a program that runs and one that doesn't.

Processing tries to tell you where you've made mistakes and to guess what the mistake is. When you click the Run button, if there are grammar (syntax) problems with your code (we call them *bugs*), then the Message Area turns red and Processing tries to highlight the line of code that it suspects as the problem. The line of code with the bug is often one line above or below the highlighted line, though in some cases, it's nowhere close. The text in the Message Area tries to be helpful and suggests the potential problem, but sometimes the message is too cryptic to understand. For a beginner, these error messages can be frustrating. Understand that Processing is a simple piece of software that's trying to be helpful, but it has a limited knowledge of what you're trying to do.

Long error messages are printed to the Console in more detail, and sometimes scrolling through that text can offer a hint. Additionally, Processing can find only one bug at a time. If your program has many bugs, you'll need to keep running the program and fix them one at a time.

Please read and reread the following suggestions carefully to help you write clean code.

Functions and Parameters

Programs are composed of many small parts, which are grouped together to make larger structures. We have a similar system in English: words are grouped into phrases, which are combined to make sentences, which are combined to create paragraphs. The idea is the same in code, but the small parts have different names and behave differently. *Functions* and *parameters* are two important parts. Functions are the basic building blocks of a Processing program. Parameters are values that define how the function behaves.

Consider a function like **background()**. Like the name suggests, it's used to set the background color of the Display Window. The function has three parameters that define the color. These numbers define the red, green, and blue components of the color to define the composite color. For example, the following code draws a blue background:

```
background(51, 102, 153);
```

Look carefully at this single line of code. The key details are the parentheses after the function name that enclose the numbers, the commas between each number, and the semicolon at the end of the line. The semicolon is used like a period. It signifies that one statement is over so the computer can look for the start of the next. All of these parts need to be there for the code to run. Compare the preceding example line to these three broken versions of the same line:

```
background 51, 102, 153; // Error! Missing the parentheses
background(51 102, 153); // Error! Missing a comma
background(51, 102, 153) // Error! Missing the semicolon
```

The computer is very unforgiving about even the smallest omission or deviation from what it's expecting. If you remember these parts, you'll have fewer bugs. But if you forget to type them, which we all do, it's not a problem. Processing will alert you about the problem, and when it's fixed, the program will run well.

Color Coding

The Processing environment color-codes different parts of each program. Words that are a part of Processing are drawn as blue and orange to distinguish them from the parts of the program that you invent. The words that are unique to your program, such as your variable and function names, are drawn in black. Basic symbols such as (), \[\], and > are also black.

Comments

Comments are notes that you write to yourself (or other people) inside the code. You should use them to clarify what the code is doing in plain language and provide additional information such as the title and author of the program. A comment starts with two forward slashes (//) and continues until the end of the line:

```
// This is a one-line comment
```

You can make a multiple-line comment by starting with /* and ending with */. For instance:

```
/* This comment
   continues for more
   than one line
*/
```

When a comment is correctly typed, the color of the text will turn gray. The entire commented area turns gray so you can clearly see where it begins and ends.

Uppercase and Lowercase

Processing distinguishes uppercase letters from lowercase letters and therefore reads "Hello" as a distinct word from "hello". If you're trying to draw a rectangle with the rect() function and you write rect(), the code won't run. You can see if Processing recognizes your intended code by checking the color of the text.

Style

Processing is flexible about how much space is used to format your code. Processing doesn't care if you write:

```
rect(50, 20, 30, 40);
```

or:

```
rect (50,20,30,40);
```

or:

```
rect     (          50,20,
   30,    40)              ;
```

However, it's in your best interest to make the code easy to read. This becomes especially important as the code grows in length. Clean formatting makes the structure of the code immediately legible, and sloppy formatting often obscures problems. Get into the habit of writing clean code. There are many different ways to format the code well, and the way you choose to space things is a personal preference.

Console

The Console is the bottom area of the Processing Development Environment. You can write messages to the Console with the `println()` function. For example, the following code prints a message followed by the current time:

```
println("Hello, Processing.");
println("The time is " + hour() + ":" + minute());
```

The Console is essential to seeing what is happening inside of your programs while they run. It's used to print the value of variables so you can track them, to confirm if events are happening, and to determine where a program is having a problem.

One Step at a Time

We recommend writing a few lines of code at a time and running the code frequently to make sure that bugs don't accumulate without your knowledge. Every ambitious program is written one line at a time. Break your project into simpler subprojects and complete them one at a time so that you can have many small successes, rather than a swarm of bugs. If you have a bug, try to isolate the area of the code where you think the problem lies. Try to think of fixing bugs as solving a mystery or puzzle. If you get stuck or frustrated, take a break to clear your head or ask a friend for help. Sometimes, the answer is right under your nose but requires a second opinion to make it clear.

B/Data Types

There are different categories of data. For instance, think about the data on an ID card. The card has numbers to store weight, height, date of birth, street address, and postal code. It has words to store a person's name and city. There's also image data (a photo) and often an organ donor choice, which is a yes/no decision. In Processing, we have different data types to store each kind of data. Each of the following types is explained in more detail elsewhere in the book, but this is a summary:

Name	Description	Range of values
int	Integers (whole numbers)	−2,147,483,648 to 2,147,483,647
float	Floating-point values	−3.40282347E+38 to 3.40282347E+38
boolean	Logical value	true or false
char	Single character	A–z, 0–9, and symbols
String	Sequence of characters	Any letter, word, sentence, and so on
PImage	PNG, JPG, or GIF image	N/A
PFont	Use the createFont() function or the Create Font tool to make fonts to use with Processing	N/A
PShape	SVG file	N/A

As a guideline, a **float** number has about four digits of accuracy after the decimal point. If you're counting or taking small steps, you should use an **int** value to take the steps, and then perhaps scale it by a **float** if necessary when putting it to use.

There are more data types than those mentioned here, but these are the most useful for the work typically made with Processing. In fact, as mentioned in Chapter 10, there are infinite types of data, because every new class is a different data type.

C/Order of Operations

When mathematical calculations are performed in a program, each operation takes place according to a pre-specified order. This *order of operations* ensures that the code is run the same way every time. This is no different from arithmetic or algebra, but programming has other operators that are less familiar.

In the following table, the operators on the top are run before those on the bottom—therefore, an operation inside parentheses will run first and an assignment will run last:

Name	Symbol	Examples
Parentheses	()	a * (b + c)
Postfix, Unary	++ -- !	a++ --b !c
Multiplicative	* / %	a * b
Additive	+ -	a + b
Relational	> < <= >=	if (a > b)
Equality	== !=	if (a == b)
Logical AND	&&	if (mousePressed && (a > b))
Logical OR	||	if (mousePressed || (a > b))
Assignment	= += -= *= /= %=	a = 44

D/Variable Scope

The rule of variable scope is defined simply: a variable created inside a block (code enclosed within braces: { and }) exists only inside that block. This means that a variable created inside setup() can be used only within the setup() block, and likewise, a variable declared inside draw() can be used only inside the draw() block. The exception to this rule is a variable declared outside of setup() and draw(). These variables can be used in both setup() and draw() (or inside any other function that you create). Think of the area outside of setup() and draw() as an implied code block. We call these variables *global variables*, because they can be used anywhere within the program. We call a variable that is used only within a single block a *local variable*. Following are a couple of code examples that further explain the concept. First:

```
int i = 12;   // Declare global variable i and assign 12

void setup() {
  size(480, 320);
  int i = 24; // Declare local variable i and assign 24
  println(i); // Prints 24 to the Console
}

void draw() {
  println(i); // Prints 12 to the Console
}
```

And second:

```
void setup() {
  size(480, 320);
  int i = 24; // Declare local variable i and assign 24
}

void draw() {
  println(i); // ERROR! The variable i is local to setup()
}
```

Index

About the Authors

Casey Reas is a Professor at UCLA's Department of Design Media Arts. His software, prints, and installations have been featured in numerous solo and group exhibitions at museums and galleries in the United States, Europe, and Asia. Casey cofounded Processing with Ben Fry in 2001.

Ben Fry is principal of Fathom, a design and software consultancy located in Boston. He received his PhD from the Aesthetics + Computation Group at the MIT Media Laboratory, where his research focused on combining fields such as computer science, statistics, graphic design, and data visualization as a means for understanding information. Ben cofounded Processing with Casey Reas in 2001.

Colophon

The body typeface is Benton Sans designed by Tobias Frere-Jones and Cyrus Highsmith. The code font is TheSansMono Condensed Regular by Luc(as) de Groot. The display typeface is Serifa designed by Adrian Frutiger.

CPSIA information can be obtained
at www.ICGtesting.com
Printed in the USA
LVHW081822180819
628046LV00004B/53/P

9 781457 187087